W9-BCZ-926

GOODS AND VIRTUES

Michael Slote

CLARENDON PRESS · OXFORD
1983

Oxford University Press, Walton Street, Oxford OX2 6DP
London Glasgow New York Toronto
Delhi Bombay Calcutta Madras Karachi
Kuala Lumpur Singapore Hong Kong Tokyo
Nairobi Dar es Salaam Cape Town
Melbourne Auckland
and associated companies in
Beirut Berlin Ibadan Mexico City Nicosia

OXFORD *is a trade mark of Oxford University Press*

Published in the United States by
Oxford University Press, New York

British Library Cataloguing in Publication Data

Slote, Michael
Goods and virtues
1. Virtue
I. Title
179'.9 BJ1521
ISBN 0-19-824707-9

Library of Congress Cataloging in Publication Data

Slote, Michael A.
Goods and virtues.
Includes index.
1. Ethics. 2. Good and evil. 3. Virtues. I. Title.
BJ1012.S517 1983 179'.9 83-6567
ISBN 0-19-824707-9

Typeset by Joshua Associates, Oxford

Printed in Great Britain by
The Thetford Press Ltd., Thetford, Norfolk

For
Cressida and Nathaniel

Acknowledgements

Like most books in philosophy, *Goods and Virtues* has accumulated a number of debts on the path to publication. I would like to thank Thomas Nagel, John Rawls, Thomas Scanlon Jr., and, especially, Samuel Scheffler for their comments on various parts of the book; and also to acknowledge a general indebtedness to the writings of Bernard Williams. In addition, thanks are due to *The Pacific Philosophical Quarterly* for permission to use an expanded version of 'Goods and Lives' (October 1982, pp. 311–26) as the first chapter of the present book; as well as to Olive Murtagh for help with typing, proof-reading, and compiling the index.

M. A. S.

Contents

Introduction

In what follows I shall be exploring our common understanding of certain aspects of human well-being and human virtue. Judged by that standard, it will turn out that philosophers' descriptions of personal good and of virtue are in some respects not qualified enough, but that in other respects, philosophers have placed unwarranted restrictions on what counts as a personal good or virtue through reliance on *a priori* ethical theses that cannot be sustained when we confront them with the actual phenomena of the moral life.

Thus in the first half of this book I shall explain why human virtue and personal good cannot always be understood as absolutely as philosophers have tended to do, but in the second half I hope to show that certain philosophically motivated limitations on human good and excellence are unjustified and unnecessary. To that extent the unity of the present work is no greater, but also no less, than what one would expect to find, say, in an article that attempted to show that a certain philosophical analysis was in some respects too broad and in others too narrow. I shall be offering a corrective to certain prevalent views and approaches, rather than a comprehensive picture either of virtue or of personal good (or of their interrelations). And the reason for dealing with both these topics in one place is that previous discussions of virtue and personal good exhibit in rather similar ways the failures to qualify and the excessive *a priori* restrictions to which ethical thinking appears to be liable.

Discussions in ethics often proceed as if it were an all-or-nothing question whether certain things count as human goods or certain personal traits as virtues; but in Chapters 1–3, I hope to show that this assumption involves an oversimplification and that a more fine-grained approach to these topics is sometimes needed. Some things are unqualified goods or virtues, but a complete picture of human good and excellence—and of our own deep-seated attitudes to these matters—requires us to pay attention to things whose status as personal goods or virtues must be qualified in various ways.

Thus in Chapter 1, I discuss two forms of 'time preference', both a pure sort of time preference according to which the sheer timing of a good can make a difference to how much that good contributes to the overall goodness of a life, and a 'time of life' kind of time preference which regards the characteristic goals and successes of certain life periods (e.g., adulthood or the prime of life) as having greater significance for life overall than those of certain others (e.g., childhood). Time preference in one or another or both of these forms has been opposed by such philosophers as Rawls, Sidgwick, Sen, Nagel, and Fried. But I hope to show by appeal to our own intuitive reaction to various examples that neither sort of time preference is as irrational as almost all philosophers assume. (Bernard Williams is an exception, and I shall compare and contrast his views with the approach taken here.) The case against time of life time preference will lead us to the idea that certain personal goods are goods only in relation to a particular time of life and that certain reasons for action, because likewise relative to a time of life, are incapable of transmitting their force transtemporally. And I shall argue in particular against Thomas Nagel's view that a proper sense of life's unity requires us to think of all reasons for action as capable of transmitting their force throughout a life.

Chapter 2 continues the discussion of relativity with an attempt to show that certain good traits of character, certain virtues, are virtues not in any absolute sense, but only in relation to certain periods of life or to certain possible worlds. The chapter begins with a discussion of recent work on the subject of life plans. Philosophers like Rawls, David Richards, Charles Fried, and John Cooper have all defended the importance for ethical theory of the notion of a life plan, but, without questioning the fundamental significance of the notion, I claim that these philosophers have failed to indicate some important limitations on the validity of rational life-planfulness. I argue that the having of a general life plan is in many circumstances essentially counterproductive, and, contrary to the claims of Rawls, Richards, *et al.*, it is maintained that some of life's most basic goods are not reasonably treated as goals within a life plan. I then point out that however valuable some (perhaps limited) form of life-planfulness is in adulthood, the possession of a worked-out life plan is undesirable and positively counterproductive in childhood; and this leads to the view that while life-planfulness may be a virtue in relation to adulthood, it is an anti-virtue with respect to childhood. Unlike such virtues as courage, temperance, and wisdom, which are seen always and essentially to be virtues, life-planfulness should be regarded as a virtue only in relation to adulthood, and is thus not an absolute virtue, not a virtue *tout court*. And similar arguments are then deployed towards similar conclusions regarding such (supposed) virtues as prudence, innocence, and faith.

In Chapter 3, I take up another sort of qualification that can usefully be incorporated in discussions both of personal good and of virtue. Some things appear to be goods whenever they occur, but others seem to count as such only when accompanied by some other (or certain other) goods; and in an exactly parallel way certain virtues (e.g., conscientiousness) can be regarded *as* virtues, *as* good things, only when certain other virtues (e.g., basic decency) underlie them. Such goods and virtues I call dependent goods and dependent virtues, and an attempt is made to show by a variety of examples just how useful and necessary such a notion is to our understanding of ethical phenomena and, most particularly, to our most deeply held ideas about social justice. Rawls has defended what he calls the primacy of justice as a virtue of social institutions on the basis of what he takes to be the overriding force of considerations of justice in the proper regulation of social institutions. But in Chapter 3, I suggest another way in which we may try to account for our sense of the special status or primacy of justice among the various virtues of societies, namely, by showing it to be a social virtue on whose presence all other social virtues (e.g., community and civility) depend for their value.

The presumptive facts of relativity and dependency to be discussed in the first three chapters do not, however, imply a failure of moral or ethical objectivity. Unlike more familiar forms of relativity, we shall see that they do not entail a relativism that conceives the qualified goods and virtues we shall discuss as created or affected by human choice or belief. But in the main I shall avoid considering meta-ethical or epistemological issues—and the broader topics of semantic theory—and simply assume that we may properly enquire about what is objectively right, good, or admirable.

Our ordinary moral thinking encourages us, then, to qualify the status of certain things as personal goods or virtues. But, on the other hand, I think various familiar attempts to argue against ordinary conceptions of what is a virtue or good for people on the basis of *a priori* philosophical assumptions are misguided. And in Chapters 4–6, I shall attempt to show not only that moral considerations operate as less of a constraint on what counts as admirable or as a virtue than philosophers generally recognize, but that moral requirements and ideals of excellence themselves, in turn, do not limit what counts as a personal good, as an element in personal well-being, in the way they are so often thought to do.

Chapter 4, in particular, seeks to establish the possibility of contra-moral virtues, of virtues that run counter to morality. There is a widely held 'overridingness thesis', according to which (very roughly) moral judgements override all opposing considerations that may occur to moral agents. I attempt, again largely by example, to show the weakness of this general thesis and proceed from there to point out some cases where a character trait that inherently involves a tendency to wrong action can none the less be regarded as admirable, and as a virtue. In the recent literature of utilitarianism there has been much discussion of how certain character traits or motives may have a (larger) moral justification despite involving a tendency to wrongdoing in particular cases. But the examples I shall mention are ones in which the trait of character that involves a tendency to wrong actions in some cases cannot be given any larger or overall moral justification, and such traits thus arguably represent something that can properly be called admirable immorality. I shall claim that the possibility of admirable immorality does not require, or depend upon, a dismissive attitude towards common morality, and attempt to defend the notion against a wide variety of possible objections.

Chapter 5 discusses the limits that virtue or morality has sometimes been said to place on what counts as a personal good for someone. Without going so far as to claim, with Plato, that the life of virtue can be shown to be to the advantage of everyone, even the immoralist, John McDowell has recently argued that the wealth, power, and pleasure that a virtuous individual has to forgo in the name of virtue do not count as personal goods from his distinctive standpoint: the virtuous individual, at least, loses nothing by being virtuous. This view is defended on the basis of two assumptions that have independent plausibility and appeal: first, that the virtuous individual can be thought of as having no reason whatever (from his distinctive standpoint) to seek a supposed good that he knows can be achieved only immorally or dishonourably; second, that if something counts as a good, one has at least a prima-facie reason to want to obtain (attain) it. And I argue that the first of these assumptions is an important moral/psychological insight, but that the second, which McDowell and a host of others have treated as *a priori* valid and 'definitional', is inadequate as a description of the way we actually think about reasons and goods. It turns out that such things as pleasure, wealth, etc. count as personal goods for the virtuous individual, even in those circumstances where he must forgo them, and the view is briefly defended that such things as wealth, success and pleasure (even the pleasure of a sadist or drug addict) may always count as personally good things. (Personal goods are distinguished, however, from 'intrinsic goods', from what is a good thing from some impersonal or disinterested standpoint.)

Chapter 6 then takes up and criticizes another, quite different attempt to limit (common-sense notions of) human good through philosophical argument. The Stoics maintained, on the basis of an ideal of self-sufficiency (*autarkeia*), that such things as love, friendship, wealth, success, and pleasure are not really good (for us). But I attempt to show that Stoicism is forced to make unrealistic assumptions about what human beings are capable of, and go on to question its assumption that human good must be measured by the standard set by an ideal or higher form of being not subject to various human limitations. What is good for us, even what is good in general, is not necessarily determined by what a higher or (more) perfect sort of being would need or want.

Chapters 4–6 have, it would seem, a common thrust. Once we reject the attempts discussed in those chapters to set limits to virtue and personal good, I believe we arrive at a more realistic and (in the best sense) more worldly view of ethical phenomena. For we can then see that moral considerations are not so uniquely and all-powerfully important as is generally imagined, that the connection between virtue and what counts as good (for a virtuous individual) is not so tight as we would sometimes like to think, and that human well-being cannot be assured by an other-world-liness which, by modelling our good on the needs and interests of ideal, non-earthly beings, repudiates any good that involves risk or our distinctively human needfulness. But having said this much about the themes and discussion of different parts of this book, let me now begin by considering the role that the timing of goods plays in determining the overall goodness of lives.

1

Goods and Lives

The fact that goods occur at one time (of life) rather than another is generally held to make no difference to the overall goodness of lives or to the existence of reasons for action, and this rejection of 'time preference' can be found across a wide spectrum of moral theorists. But in this first chapter I shall argue that time preference is not the atypical or irrational phenomenon that so many otherwise divergent theorists assume and that our ordinary thinking quite naturally ascribes unequal importance to different periods of life.

The discussion will focus on some examples that have largely escaped the notice of those opposed to time preference, examples that I believe cannot be adequately explained without making some new conceptual distinctions about human good and using the philosophically rather neglected concept of 'the prime of life'. And the examples chosen will also enable us to question the equal importance of all time periods and the automatic transtemporal influence of reasons for action, without having to presuppose the sort of intrapersonal dissociation or disunity whose invidiousness has so largely motivated opposition to any sort of time preference. Our defence of time preference will be different, moreover, from what the most notable recent defender of the notion, Bernard Williams, has said on its behalf; and since the repudiation of temporal inequality seems to be an important underlying assumption both of Utilitarianism and of such major anti-Utilitarians as Rawls and Nagel, it will also be important to see how our particular arguments against that assumption affect the various theories that have hitherto partly based themselves upon it. But first I would like to give some indication of the breadth and depth of its appeal.

I

Let us begin with the maximizers. Utilitarians seem to take the denial of time preference as a practically self-evident condition on any maximizing conception of human good or rational action. Sidgwick, for example, holds that the maximization of personal welfare over time pre-supposes the equal treatment of all the times in a person's life in much the same way that the maximization of social welfare (or the welfare of all sentient beings) requires us to accord equal weight to the welfare of every individual.[1] And for all their other disagreements, John Rawls follows the Utilitarians in presenting a maximizing conception of the goodness of single lives and of egoistically rational life-planning that treats all time periods in a single life equally: the timing of goods being important only as a means to maximizing individual well-being over time.[2] And he also argues (in a manner at least partially anticipated by Sidgwick)[3] that time preference involves 'not viewing all moments as equally parts of one's life'.[4] Of course, Rawls criticizes Utilitarianism for extending to society the (maximizing) principle of rational choice appropriate for one man.[5] He holds that the Utilitarian's conception of social (moral) choice mistakenly treats social groups as organic superindividuals, and that a proper sense of the distinctness of individuals within any social group entails a preference for his own non-maximizing principles of social justice. But he shares the Utilitarian maximizing conception of individual good and the rejection of time preference that, as we have noted, seems naturally to

[1] *The Methods of Ethics*, London: Macmillan, 1907, seventh edition, pp. 381 f.

[2] See *A Theory of Justice*, Cambridge: Harvard University Press, 1971, secs. 45, 64.

[3] Op. cit., pp. 418 f. [4] Op. cit., p. 295. [5] Ibid., pp. 26 f.

accompany it, and others influenced by Rawls and likewise opposed to Utilitarian moral theory have also insisted on the equal treatment of times within single lives. Thus Thomas Nagel in *The Possibility of Altruism* has attempted to show that all periods of an individual's life must play an equal role in the origination and transmission of reasons for action, and like Rawls (though more systematically) he attempts to derive the equality of different times from a proper metaphysical conception of them as forming the equally real parts of a single finite life.[6]

Moreover, the idea that periods in a single life are to be treated equally is not simply a presupposition or natural adjunct of the idea of maximization. For it can also be found among philosophers who explicitly doubt or deny the validity of maximization as a principle of rational choice or individual good. Thus Amartya Sen has recently pointed out that the tragedy of King Lear's fate is not thought to be effectively blunted by any facts about how fortunate he was in earlier years, and he has (tentatively) concluded that the fortunateness of a life cannot simply depend on how much good (or good-minus-evil) it contains.[7] Sen suggests that what is wrong with Lear's life is a lack of reasonable equality, with respect to goods and evils, between the different periods of that life. In the determination of intertemporal human good—the goodness of lives—he thinks the roughly equal intertemporal distribution of goods carries weight quite independently of good-maximizing considerations.

However, in saying all this, Sen is not questioning the maximizers' assumption of temporal equality, but rather giving and accepting a rather strong interpretation of that notion. Consider the parallel interpersonal case. A Utilitarian may claim to give the fullest valid expression to the

[6] *The Possibility of Altruism*, Oxford: Clarendon Press, 1970, Ch. 8 (esp. p. 60).

[7] 'Utilitarianism and Welfarism', *Journal of Philosophy* LXXVI, 1979, 470 f. Sen abstracts from literary-critical considerations that might suggest that Lear, through suffering, attained something better at the end of his life than anything in his previous life, that Lear's life ended better than it began.

idea of human equality by allowing each person to 'count for one' in determining what course of action or policy maximizes social (or overall human) welfare, but an advocate of an equal distribution of benefits and burdens in society will hold that his own emphasis on the independent value of equality gives a far better expression to what we cherish in the idea of interpersonal equality. The Utilitarian and the egalitarian may thus both claim to speak in the name of true and proper equality, and so can be thought of as presenting essentially competing conceptions (or expressions or interpretations) of a single underlying concept (or idea).[8] By the same token, intrapersonal maximizers may claim that indifference to when given satisfactions or goods occur in the process of calculating total welfare over time is the most proper expression of the idea of intrapersonal temporal equality, but an egalitarian might want to say that the idea is better expressed by placing an independent value on the intertemporal equality of goods within single lives. And although it is no part of our business here to adjudicate this dispute, I do want to emphasize that Sen's temporal egalitarianism seems to leave no more room for time preference than Sidgwick, Rawls, and Nagel do.

However, one may also oppose maximizing conceptions of rational choice and individual good through a more articulated conception of the value of intertemporal equality than Sen presents. Charles Fried rejects an egalitarianism of sheer temporal moments (or intervals) in favour of (what seems to be) an egalitarianism regarding the various stages or periods of the human life cycle. Childhood, youth, maturity, etc., have, he says, their own characteristic needs and projects (in addition to those that persist across such periods) and a rational plan for a good life involves not only a maximizing of goods realized but

[8] The above exposition benefits from ideas of Rawls (op. cit., sec. 1) and of Ronald Dworkin (*Taking Rights Seriously*, Cambridge: Harvard, 1978, esp. p. 180).

the giving of 'richness and realization to each period of life'.[9] Now the fact (so greatly emphasized by Fried) that time periods within single lives are plausibly carved out or articulated by natural and socially influenced facts about the typical human life cycle will turn out to be highly relevant to our own defence of the intrapersonal inequality of times, but Fried, it should be clear, makes no such use of the notion: like the other philosophers we have discussed, he seems to be suggesting that different (properly articulated) times of life are of (roughly) equal importance in determining the goodness of lives.

Rawls, Nagel, Sen, Fried, and even Sidgwick all lay stress on the need to test their views against reflective common-sense judgements of value and ideas of the self and human life,[10] and in rejecting time preference as irrational they seem to assume they are giving expression only to what is already enshrined in considered opinion about what makes lives fortunate and what gives people reasons for acting. But I believe and shall argue, on the contrary, that we typically and naturally think of some times of life as more important than others, and that this conclusion has important implications for our notions of personal good.

II

Human life seems, as I have said, to possess a natural, though socially influenced, development of different times or stages of life. Whether we speak, with Shakespeare, of 'seven ages of man' or, with Erikson,[11] of the eight stages of the 'life cycle', it seems plausible to suppose that a proper understanding of human life as human beings themselves see it requires a division into different periods that

[9] See *An Anatomy of Values*, Cambridge, Harvard, 1970, pp. 170–6.

[10] See Rawls, op. cit., p. 19 f. and elsewhere; Nagel, op. cit., pp. 80, 126, 144 f. (though contrast p. 82); Sen, op. cit., pp. 470 f.; Fried, op. cit., Ch. 10, *passim*; and Sidgwick, op. cit., Book III, Chs. XI and XIII.

[11] See *Childhood and Society*, N.Y.: Norton, 1963, 2nd ed., pp. 271 ff.

reflects not an indifferent mathematical conception of infinite successive moments but rather the definition and contour that common sense attributes to our lives by speaking, e.g., of youth, adulthood, and old age.

But I believe that such a division into 'times of life' tends to be accompanied, in most of us, by a sense of the greater importance or significance of certain times of life in comparison with others, and what I first want particularly to stress is the lesser seriousness with which we regard the successes and misfortunes of childhood (including adolescence) when considering, in the rough and ready way we sometimes do, how fortunate someone has been in life. I think we have a definite tendency to discount youthful misfortune or success that can be seen, for example, in what we think and say about someone who won all the prizes and captained all the teams in school, but whose later life seems dull or unfortunate by comparison. Hearing such things about people one knew in childhood, one may knowingly shake one's head: and what is peculiarly sad about such cases can be expressed, I think, in the thought that schoolboy (or schoolgirl) glories cannot compensate for (cannot begin to make up for) what happens later in life. But by the same token schoolboy misfortunes are also largely discounted. A statesman known to have led a very happy and successful adult life may be discovered to have had a miserable childhood, but unless we imagine that that embittered his adulthood in ways not immediately obvious from other biographical facts, I don't think our discovery will make us wonder whether we haven't been over hasty in supposing the man (or his life) to have been fortunate, enviable. Within a very wide range, the facts of childhood simply don't enter with any great weight into our estimation of the (relative) goodness of total lives.

In a way, our treatment of childhood (though, as we shall see, not exclusively of childhood) is interestingly similar to the way we regard what happens in dreams, Proust tells us (roughly) that we do not reckon the sufferings and pleasures of our dreams among the actual goods and evils of our lives.[12] And it would indeed seem that we have no tendency to consider someone's fortunateness or unfortunateness to be in any way a function of the quality of his dreams during his life. (We are not likely to withhold a judgement on how well someone has lived until we find out how he spent his nightly hours of sleep.) And just as dreams are discounted except as they affect (the waking portions of) our lives, what happens in childhood principally affects our view of total lives through the effects that childhood success or failure are supposed to have on mature individuals. Thus in cases where an unhappy schoolboy career is followed by (or, as we sometimes like to think, helps to bring about) happy mature years, we think of the later years as compensating for the childhood misery, even as wiping the slate clean, and I believe that Rawls, Sidgwick, and others who have assumed the equal status of all times of life have not taken this sort of common judgement sufficiently into account.

To make theoretical sense of the tendencies of thought we have just described and do so without having to attribute undue irrationality to common patterns of human thinking, we need to make some new distinctions. In order to see why, let us first consider whether our relative indifference to certain aspects of childhood cannot be plausibly understood in terms of the already familiar distinctions between rational and irrational desires and between a person's mere preferences (desires) and his values.[13] Can we, for example, explain our attitudes toward childhood on analogy with the unwilling drug addict or sadist who places no value on satisfying the cravings he finds himself subject to? The (or one kind of) unwilling addict would rather not have his (irrational) cravings; having them, he would rather not give in to

[12] *Cities of the Plain*, N.Y.:Random House, 1970, p. 272.

[13] Cf. Rawls, op. cit., pp. 416 ff.; Nagel, op. cit., pp. 40 ff., and Gary Watson, 'Free Agency', *Journal of Philosophy* LXXII, 1975, 205-20.

them (satisfy them); and having satisfied them, he regrets having done so. And by the same token, an un-willing or reformed sadist may wish he had somehow failed to satisfy certain (former) sadistic impulses, because, again, he (now) finds them loathsome and irrational and places no *value* on satisfying them.[14]

Perhaps such examples give us everything we need to understand the way we discount what happens in childhood. Perhaps children eagerly seek certain goals—membership of the school team, scout merit badges—and simply learn later, as adults, that such things are not worth pursuing, not really valuable. But competition for merit badges, school teams, high marks is quite characteristic of childhood and adolescence, and if we were really to press home the analogy with sadism and addiction, then we would have to claim that childhood was an irrational period of human life, characteristically fraught with the desire for and valuing of things one shouldn't value. We would have to say that the miseries, elations, and disappointments connected with success and failure in typical childhood aims were perverse and/or irrational in the way we commonly think of sadism or addiction.

All this would be utterly implausible. And that is no doubt the reason why philosophers like Rawls and Nagel, who make use of the distinction between rational and irrational desires (or between values and mere preferences) and recognize its application to examples like addiction or sadism, never use that distinction to differentiate the characteristic goals and strivings of childhood from those of other periods of human life. (Indeed to do so would involve them in what, as we shall see, they themselves would have to count as time preference.) Our problem, then, is to understand how we (adults) can discount typical childhood strivings, successes, and disappointments without being unfair to childhood in the way that

[14] I believe I here disagree with what Phillip Bricker ('Prudence', *Journal of Philosophy* LXXVII, 1980, p. 383) says about similar cases.

the strict analogy with sadism and addiction would force us to be.

We think that the cravings of (unwilling) drug addicts make them irrational and becloud their judgement. They are momentarily weak or disabled; and there is, over time, something wrong with them and unfortunate about them. But we regard most schoolboy interests and goals in none of these ways, and although the reformed, cured, or temporarily clear-headed addict or sadist will typically repudiate and be repelled by his own actions and desires (and their attendant satisfactions), that is not our attitude to childhood. We may not take childhood's characteristic goals and disappointments very seriously, but neither is our attitude one of revulsion and repudiation. Instead, we are rather tolerant of schoolboy strivings and interests and find them appropriate to, and acceptable at, the period of life in which they characteristically occur. And this latter fact may actually offer a clue about how to distinguish the childhood goals we subsequently take so little seriously, from sheer irrationalities.

The desire for honour-roll marks and the like are appropriate to their time of life—to *some* time of life—in a way that addiction and sadism are not, so even if an adult doesn't take such desires seriously and discounts the value of their objects, he may still be able to say that those objects—honour-roll marks, captaincy of the basketball team—are valuable at or for the time of life when they occur. He may say, in other words, that such things have value for, or in, childhood, but not value *ueberhaupt*, i.e., not value from the perspective of human life as a whole.[15] And this would explain his

[15] Talk about what is good in, or from the perspective of, childhood is not reducible to the notion of being *thought* good by children, because such a reduction would, once again, fail to distinguish childhood goals from irrational addictions. We really require another argument place in order to describe (the good of) those goals—see below. It is also worth pointing out how the notion of childhood-relative goods gives rise to other locutions (some of which we

discounting yet accepting attitude towards characteristic childhood successes and failures; it would enable us to understand why so many of us think such successes and failures make such a negligible contribution to the overall value or unfortunateness of a human life and yet find them appropriate to childhood and adolescence rather than, like sadism or addiction, a cause for despair or regret.

Moreover, the distinction between period-relative and overall human goods is useful not only in differentiating characteristic aspects of childhood from truly irrational phenomena, but in making some intuitive distinctions within childhood itself. A twelve-year-old who clings to his mother whenever company comes, may himself feel too old for such things and consequently ascribe no value to his success in clinging on various company occasions. And it seems plausible to liken the clinging to a kind of unwilling addiction and say that, unlike getting good marks or being a team captain, the clinging has no value for the twelve-year-old, (even) at, or for, his own time of life. We are accustomed to the idea that goods (benefits, misfortunes, etc.) are often goods for particular individuals rather than free-floating and 'objective'. But the sentence before last suggests that our distinction between overall and period-relative goods requires in some cases a double relativity: not merely of goods to individuals, but of goods to individuals at times of life.[16] Such double relativity may not be familiar and, for that reason, its introduction must be properly motivated. But I hope the above discussion gives some indication of why efforts to formulate a plausible (and non-debunking) account of our actual attitudes towards childhood naturally reach out for

shall be using in what follows). For example, what is good only in childhood is naturally described as being good *for children* (as opposed to being good for persons, for the persons who were such children).

[16] Nagel, op. cit., argues against the possibility of independent person-relative reasons for actions, but in other writings seems to want to allow for the possibility of person-relative (subjective) goods and reasons for action. See, e.g., *Mortal Questions*, Cambridge: C.U.P., 1979, pp. 2, 132 ff., 203.

the distinctions we have just introduced, in addition to the more familiar ones I have mentioned.[17]

We have thus far focused upon certain attitudes towards childhood and upon the idea of childhood-relative goods that they seem to imply, but I want to make room for the possibility that similar period-relativity may be helpful in understanding attitudes towards other times (periods) of life. For example, I have somewhere read that the Romans regarded the study of philosophy as a pursuit suitable for a young man, but not for a fully mature individual, and even if my particular readers are likely to disagree with this evaluation, it may be that in order to understand and best express the Roman perspective on philosophy, we must again speak of period-relative goods and attribute to them the view that philosophy has its appropriate place and value for a young man (for the period of young manhood), but is not to be taken fully seriously from a more mature perspective. Or, to take a less uncomfortable illustration, consider the (or a) common attitude taken to what happens during senescence or old age. Our view of a person—once a successful architect, academic, or politician—who while in retirement during his declining years concentrates on winning senior citizen shuffleboard tournaments, may be similar to what we think of childhood goals and strivings. His victories will count negligibly or not at all towards making his overall life seem happy or fortunate, because we tend not to take such successes very seriously, but they may none the less be taken as goods relative to his particular time of life. (A senior citizen may be fortunate overall to win if he in particular would otherwise become embittered or disconsolate, but that is another

[17] When St. Paul speaks of putting aside childish things, he also seems to allow that understanding and speaking as a child are all right in childhood. Our notion of the childish thus seems to contain an ambiguity, or bipolarity, between the childish as what is irrational-in-the-manner-of-a-child and the childish as what is appropriate to children only. This is close to the distinction we have been urging above in connection with childhood vs. addiction, and it supports the idea of period-relative goods for reasons already mentioned.

matter, a matter of means and ends, and we shall in any event find it sad that he takes senior citizen shuffleboard so deadly seriously.)

Of course, we can use slightly different language in expressing our attitude from what would be appropriate in speaking of childhood. We may say, for instance, that it is sad for a former architect or what-have-you to be reduced to putting so much emphasis on shuffleboard tournaments, and we would not similarly speak of a child's being *reduced* to the desires that seem distinctive of childhood. But this need not mean that we regard old-age strivings as irrational in the manner of addictions. Senescence itself may be sad and regrettable, but shuffleboard competitions can none the less seem appropriate as a focus of that declining time of life. They may be the best thing one can do under the circumstances, whereas it is not usually thought appropriate or best that someone in the unfortunate situation of an addict should keep satisfying his addictive cravings. So our talk of someone being 'reduced' to certain activities may indicate only that the person in question is in a state of decline relative to some earlier period, and we can thus explain why we do not use this locution of developing children, while holding that senescent shuffleboard victories are period-relative goods in the sense of our earlier, childhood examples.[18]

I am by no means claiming that all, or even most, of the satisfactions of senescence, or of childhood, are merely (period-)relative goods. But the examples we have mentioned do suggest that some of the principal goals, disappointments, successes, and satisfactions, characteristic of certain life periods, have value only relative to those periods and make a rather negligible contribution to what seems to matter most in a total human life. On the other

[18] Incidentally, we do not regard childhood strivings and goals as appropriate merely because we regard them as means to what we take more seriously in adult life. Often we don't know—have no particular theory about—whether a given appropriate-seeming childhood activity will further adult values.

hand, the period known as 'the prime of life' is typically conceived as containing precisely those goals, strivings, miseries, and satisfactions, that are to be taken most seriously in human life, and is thus largely an exception to what we have been saying about childhood and senescence.[19] Indeed, the very expression 'prime of life' conveys the implication that the failures and successes of other periods are inherently less serious and less determinative of what one's life has, for better or worse, been like. And one might even say that the idea of period-relative goods is the natural result of superimposing our sense that there is a *prime* of life on the relatively neutral framework of *periods* or *times* of life.[20]

On the other hand, one might try to evade the time-period relativity of goods by arguing that childhood (or senescent) successes merely have very little (non-relative) value for life overall. But if children spend so much time and effort pursuing what are just very minor goods, then they are like adults who expend great effort on very unimportant goods or projects, e.g., like the author who (compulsively) spends day and night searching galley proofs for a single misprint that he knows to be trivial and not sense-affecting. (He may know this through having gone over the proofs of his tome earlier and having neglected to remove a printing error he remembers to have been trivial but cannot now place.) And I think such a view makes children, e.g., seem more irrational than they are warrantedly held to be. Moreover, even if we simply say that the characteristic goals and strivings of youth and old age have very little (overall) value—by comparison with those of the prime of life—we seem committed to a time preference for the period of adulthood.

Of course, children *do* have a narrower perspective than

[19] 'Prime of life' works better for what I have in mind than either 'adulthood' or 'maturity', which do not clearly enough exclude senescence.

[20] Following up the analogy with dreams, could we not perhaps say that certain satisfactions in dreams are goods relative to those dreams or to what might be called one's dream life (on analogy with one's sex life)?

adults; recognition of that fact lies behind and helps to justify our adult willingness to make judgements about the status of childhood values. But it is the appropriateness of what children do within the naturally-given narrower perspective of their time of life that differentiates them from addicts and sadists and makes what is of negligible value for life overall a substantial good in relation to childhood. The idea of goods–relativity thus understandably takes a place *alongside* judgements of absolute (or overall) value, in attempts to discern the full complexity of human good.

III

When Rawls speaks of time preference, he has in mind the preference not only for what is strictly nearer or earlier in time but also for one or another *period* or *part* of human life; and his rejection of time preference thus leads him to treat the (rational) goals and preferences characteristic of each such period as equally serious and of equal value.[21] But we have seen that the very opposite of such an attitude is reflected in the way, e.g., childhood and extreme old age are commonly regarded, and since Rawls himself seeks (*ceteris paribus*) to be true to our ordinary considered judgements, his total rejection of any form of time preference may stand in need of qualification. Of

[21] Op. cit., pp. 420 f., for example. He seems to assume, with Fried, that the value of an activity may be assessed 'relative to its own period'. But despite the coincidence of terminology, this involves not the period-relativity of goods defended above, but rather the 'democratic' notion that the differing goals of each period give rise to overall (non-relative) life goods occurring within those periods.

Incidentally, even though the person who regards what happens in childhood as less determinative of the overall goodness of a life, may have this attitude on the basis of certain purely contingent features of childhood, in particular, its characteristic goals and strivings; this attitude may still count as a form of (impure) *time* preference. After all, someone who likes France better than Spain has what may properly be called a place preference, yet that preference is not independent of the contingent amenities of those two places.

course, what we have so far defended is not 'pure' time preference, if by that one means the favouring, say, of earlier or nearer times of life as such. Rather, it is a preference for the goals and interests characteristic of certain stages or periods of life rather than others, and these goals and interests are from a logical standpoint perhaps only contingently related to what comes earlier or later in time. So even if certain times of life in the ordinary sense count differently in determining the overall value of a life, one might still want to rule out as irrational any preference for the early or late (temporally near or far) as such.

I think, however, that even such pure time preference occupies a place in our thinking about the goodness of lives and can be found (ironically) not where Rawls is most intent on arguing against it, in any favouring of the temporally nearer or earlier, but rather in a precisely opposite preference for what comes later in life.[22] When a personal benefit or good occurs, may make a difference to how fortunate someone is (has been), quite independently of the effects of such timing in producing other good things and of the greater importance we attach to the distinctive goals and interests of certain life periods. And I believe, in particular, that what happens late in life is naturally and automatically invested with greater significance and weight in determining the goodness of lives. The point can be illustrated.

A given man may achieve political power and, once in power, do things of great value, after having been in the political wilderness throughout his earlier career. He may later die while still 'in harness' and fully possessed of his powers, at a decent old age. By contrast, another man may have a meteoric success in youth, attaining the same office as the first man and also achieving much good; but then lose power, while still young, never to regain it. Without hearing anything more, I think our natural, immediate

[22] Pure time preference for what comes later in life is defended by J. N. Findlay (*Values and Intentions*, London: Allen and Unwin, 1961, pp. 235 ff.) on the basis of arguments somewhat different from those offered here.

reaction to these examples would be that the first man was the more fortunate, and this seems to suggest a time preference for goods that come late in life.

Now one might try to explain away these reactions to the above example by showing them to arise from something other than the time preference they seem to involve. For example, almost everyone thinks it better for a person to receive an (automatically) increasing salary over the years rather than an initially high but gradually decreasing one; and one might attempt to explain *this* preference as resulting, not from any sort of time preference, but (in the manner attempted by Rawls[23]) rather as due to the superiority of pleasures of anticipation over pleasures of memory. In other words, the industrial worker whose great benefits come later will simply have greater pleasures of anticipation than any pleasures of memory one would have in the case of decreasing salary. (Why Rawls assumes that good things are more likely to be remembered with pleasure and gratitude, than with bitterness at their loss, I do not know. See George Eliot's *Daniel Deronda* for a strong statement of the opposite view.) But however well this sort of (good-maximizing) explanation suits our preference for salary increments by seniority, it cannot adequately explain the typical initial reaction to the above political example. For the person who achieves and loses high office when he is still relatively young may well hope and have reason to expect to gain power again, whereas the politician who is in the political wilderness throughout his early and middle years may easily stop expecting to gain power. And in that case, the man who succeeds late may have *fewer* of the pleasures of anticipation or hope than the one who achieves early success, and our greater estimation of the former's career presumably cannot come from any assumption to the contrary. It seems, rather, to be a

[23] Op. cit., p. 421.

matter of sheer preference for goods that come later, of our assumption, even, that a good may itself be greater for coming late rather than early in life.[24]

Such 'pure' time preference is embodied not only in our natural and (I believe) persisting reactions to particular cases, but also in the very language with which we describe how well we think people have lived. We may say that later political success can 'compensate' or 'make up' for (someone's) years in the political wilderness; but it would be an abuse of language to describe early successes as 'compensating' or 'making up' for later failures or miseries. And lest someone reply that this is merely a fact of linguistic convention, can it not be said further that the very fact that we have expressions for the way later goods can counterbalance earlier evils, but none at all for the counterbalancing of later evils by earlier goods, is a rather good indication of our common belief in the greater intrinsic importance (value or disvalue) of what comes later in life?[25]

In the light of the above, it may also at this point be worth reconsidering the example of King Lear as Sen describes it. Sen treats the example as evidence that we do not think of the goodness of lives solely in terms of maximizing criteria, but rather require good lives to embody some sort of intertemporal equality of goods. But our discussion of time preference makes another interpretation possible. The badness or tragic quality of King Lear's life may be more a function of how badly it *ends* than of any inequality

[24] Perhaps the preference for what comes late (or last) makes sense not only for total lives, but for careers within lives. Muhammad Ali is still a relatively young man, but wouldn't some people have preferred to see him end his boxing career while still heavyweight champion, the way Rocky Marciano did?

[25] Some people become embittered through early unsuccess and don't allow later success to mellow them, refusing, e.g., to see it as compensating for earlier failure. But we think it is irrational of someone if he can't or won't let later success make up for what went earlier, and the latter judgement again shows our preference for what comes later. Incidentally, given that Rawls is unequivocally opposed to time preference, I find it hard to know what to make of the sentence beginning at line 16 of page 421 of *A Theory of Justice.*

among its various periods. And this possibility can seem especially plausible when one considers how untragic a life that reversed the history of King Lear would seem. For the relative fortunateness of a life in which early misfortune was followed by a long and happy old age cannot be explained by Sen's considerations of equality: as with our earlier political example, we need to bring in time preference in order to do it justice.

We may now also be in a position more fully to account for our tendency to devalue what happens in childhood and adolescence. That tendency may be partly accounted for by the above-discussed lesser seriousness with which we regard childhood plans and interests, but some of it may be due to the separate fact that what happens early in life can be compensated for by what happens later. Both 'time of life' time preference and a purer sort of time preference may be involved. And this now more complicated picture of our attitudes towards childhood may be reinforced by the following further considerations relating to old age.

We tend to think that someone who suffers from bereavement or disease in old age is more to be pitied than someone similarly miserable in childhood (alone), but this opinion cannot entirely be due to the differential seriousness or importance of the goals and preferences of childhood and old age, since it seems in no way to depend on assuming that the person who suffers in old age is not senescent (senile). Instead, the idea seems to be that nothing can compensate for sufferings at the end of life, and that thought expresses the sort of 'pure' time preference we have just been considering. On the other hand, both the sorts of time preference we have mentioned appear to be involved in other attitudes towards old age: e.g., in the common assumption that it is better for an older person to die in harness, fully possessed of his powers, than to have a longer life of (forced) retirement subject to an increasing diminution of his powers.[26] For

[26] A point also made by Fried, op. cit.

the latter assumption seems simultaneously to express both our preference for the prime of life and the importance we place on a life's ending well. At this point, however, it is time for us to see how such double-edged time preference affects the views of some of the philosophers referred to above.

IV

As I mentioned earlier, Rawls believes that Utilitarianism can be viewed as adopting 'for society as a whole the [correct] principle of rational choice for one man [i.e. for the rational egoist]' through a mistaken conflation of all individuals into a single (super)person.[27] This 'diagnosis' has recently come under attack for its assumption that Utilitarianism makes use of, and has a distinctive need for, the metaphysics of social organicism. Derek Parfit (following Sidgwick) has pointed out that Utilitarian ethical theories are far more characteristically found with (underlaid by) atomistic/associationalist views of the individual, of the sort so frequently advocated by empiricists.[28] But if our earlier defence of time preference has any validity, then Rawls's diagnosis can also be faulted for its assumption that the Utilitarian is right to make the goodness of single lives depend in a time-indifferent way on how much good (or good-minus-evil) they contain.[29]

On the other hand, Parfit argues forcefully that radically atomistic views of individual identity would (relatively) favour Utilitarianism over deontological social moralities like Rawls's, and it may well turn out that such views are also incompatible with what we have said in favour of time preference (though we shall later have

[27] Rawls, op. cit., pp. 26 f.

[28] See 'Later Selves and Moral Principles', in A. Montefiore, ed., *Philosophy and Personal Relations*, London: Routledge and Kegan Paul, 1973, pp. 137-69. Cf. Sidgwick, op. cit., pp. 415 ff.

[29] Rawls assumes that individual good and rational choice are intimately connected. See op. cit., pp. 417-22, e.g. We shall have occasion to doubt such a connection in Chapter 5, below.

reason to defend ourselves against the precisely opposite charge that time preference fails to respect the unity of single lives). But since it is hardly clear what approach to individual identity is most likely to prove fruitful in the long run, Rawls's positive argument for his two principles of justice may in the end carry force despite a dual failure in his account of where Utilitarianism, as an ideal type, goes wrong.[30]

We have so far been concentrating on the sorts of judgements concerning individual good that are sometimes made when we consider the lives of others or stand back from our own lives and attempt to view them in a detached way. But what we have said in this connection may also be relevant to a proper understanding of the reasons for action people have in the midst of their lives. And in particular we shall see that it raises problems for the temporally egalitarian conception of reasons for action put forward by Thomas Nagel in *The Possibility of Altruism*.

According to Nagel, a proper appreciation of the equal status (equal reality) of all the times of a given person's life entails a conception of such reasons as essentially tenseless and (so?) operating across (all) the times of that life. If, as in Nagel's own example, I shall have reason to speak Italian at some future time, that gives me (some) reason now to prepare for that eventuality, e.g., by studying Italian. And if I do not study Italian and am unable to speak it when I need to, then I shall even later have reason to regret that I didn't study the language and was unable to speak at the appropriate time.[31] On Nagel's view, anyone who denies such things lacks a conception of himself as an individual all the times of whose life are equally real parts of a single life. But Nagel is also perfectly willing to allow that someone may at a given time think he has

[30] However, for doubts as to whether Rawls's argument for the two principles is really compatible with 'atomistic' theories of the individual, see Samuel Scheffler's 'Moral Independence and the Original Position', *Philosophical Studies* 35, 1979, 397–403.

[31] See Nagel, op. cit., Ch. 8.

reason to do something and yet be mistaken. An alcoholic, for example, may have reason to place a time lock on a drinks cabinet in order to take precautions against a 'later self' who will mistakenly think he has reason to open the drinks cabinet. What Nagel does not allow is that genuine later reasons for action should give rise to no earlier reason to facilitate their realization or that one should ever have no reason to regret or be glad about what one did in response to earlier reasons for action. A person with a proper sense of his life's unity over time cannot be indifferent to what he acknowledges to be a genuine reason (as opposed to a mere desire or preference) existing at some other time in his life.

In making these claims, Nagel explicitly appeals to our own sense of our lives, and what he says about the Italian-speaking example seems hard to deny; for the person who grants he will have reason to speak Italian yet thinks that gives him no reason whatever to study the language seems abnormally 'dissociated' from his own future. But if one casts one's net a bit wider, I believe one finds examples that do not sustain Nagel's general thesis. Even in 'normal' lives, there are cases where reasons do not translate across times, and some of these have in fact already been mentioned. We tend not to regret childhood failures in the way that, years later, we regret the failures of our earlier adulthood, because of our tendency to discount the characteristic goals and interests of childhood; and an adult's lack of regret, say, for his failure to become a school team captain or make the honour roll is not a sign of abnormal dissociation, but rather a perfectly ordinary and understandable way of viewing things.

I believe that Nagel's only avenue of response, at this point, would be to assimilate our indifference to our own childhood to the lack of regret that might be felt by an unwilling alcoholic who failed to get drink on a given occasion (though such a person might actually feel relieved, even elated, at missing an opportunity to drink). But earlier we saw the implausibility of such an analogy between childhood goals and irrational cravings, and we cannot therefore explain our lack of regret (gladness) about childhood failures (successes) by saying that these occurred in connection with desires we even then had no reason to pursue (and reason even to resist?). Rather the reverse, our earlier conclusion that childhood goals and interests represent childhood-relative goods suggests that those goals and interests give rise to childhood-relative reasons for action, to reasons that exist in childhood but leave no 'trace' in appropriate regret or satisfaction later on. In other words, if (many of) the things children seek are not seen as goods from the large perspective of life as a whole, then it is understandable that these things should not be the subject of regret and gladness later in life; but since the satisfaction of childhood goals and interests really has childhood-relative value, it cannot be treated like the satisfaction of (non-reason-giving) irrational cravings or sadistic impulses. Rather, what is good for children in childhood gives rise to childhood-reasons to act, to reasons children have while they are still children. But these reasons do not translate into later reasons for regret and satisfaction, because of the strictly period-relative values they represent. (Our way of telling disappointed children 'someday you'll laugh at all this' expresses, hyperbolically, the lukewarmness adults tend to feel about childhood failures and successes at the same time that it unhelpfully, and somewhat cruelly, ignores the period-relative reasons a child may have for being disappointed.)

The same points, furthermore, can be made about old age or senescence. If shuffleboard is a significant good only in (relative to) old age, then a person may in old age have reason to seek victory in senior citizen shuffleboard competitions and yet lack any reason to practise shuffleboard in his earlier years. What seems abnormal in Nagel's Italian-speaking example may here be entirely understandable. The idea of time-relative goods thus naturally leads us to question Nagel's rejection of time-dependent, or time-relative, reasons for action and his thesis that the reasons of every time are equally able to transmit their force to other times of life.

Mine, however, is not the first attempt to contest Nagel's conclusions about the action of reasons across time. Bernard Williams has recently argued that reasons for action and feeling exist essentially from the perspective of the (moving) present, so that the fact that, if one lives long enough, one will later have reason, e.g., to compete at shuffleboard may give one no present reason to practise shuffleboard (unless one has a current desire to provide for a contented old age).[32] And in the same way Williams can hold that our lukewarmness towards schoolboy successes and disappointments simply reflects the fact that most schoolboy goals fall outside the scope of our present (adult) interests and purposes.

Williams's criticisms of Nagel are essentially different from those offered here because of their reliance on a temporal perspectivalism that makes the present into the fulcrum of all reasons for action and feeling. But I see no reason to believe that Williams's approach is incompatible with our own; and it is possible that they represent two equally valid ways of questioning Nagel's general thesis about the transmission of reasons. On the other hand, Williams's emphasis on the present commits him to at least one conclusion on which our own earlier arguments appear to be neutral. Consider the case of a happy-go-lucky apolitical artist or writer who lacks all plans and ambitions for the future, but who believes that the increasingly fascist political climate of her own country is likely to radicalize her into an anti-fascist political activist who does valuable things of a sort she now admires only from afar. On Nagel's view, such a person has reason in advance to prepare for her later political activity, but Williams

[32] 'Persons, Character and Morality', in A. Rorty, ed., *The Identities of Persons*, University of California Press, 1976, esp. pp. 208 f., 216.

presumably will hold that she does not. Our own view, by contrast, seems entirely neutral on this question; for it cannot, I assume, be said that the woman's future political activity represents an inferior or time-relative value, so our previous discussion provides no way to argue *against* the transmission of reasons for action from that later time; and yet since we have said nothing about how *overall life goods* occurring at particular times translate into reasons for feeling and acting at other times, we may well wonder whether the woman has reason to prepare for her predictable (but presently unintended) subsequent activities.

For the moment, at least, I have no idea how to resolve this very delicate issue. But if it turned out that the woman does have reason to prepare, then Nagel would have the better of this particular disagreement with Williams and the latter's perspectivalism of the present would be damaged as a general thesis. Then, and perhaps only then, would we have reason to reject Williams's arguments against Nagel and rely exclusively on those developed here. On the other hand, even if we accept Williams's assumptions and conclusions, our previous discussion can account for phenomena Williams leaves untouched, because of its emphasis on the character of certain goods. It explains, for example, why certain periods seem more important than others in determining the overall goodness of lives, whereas Williams's view of reasons gives us no insight into this phenomenon (or into the above-discussed phenomena of pure time preference), because of his insistence on the separateness of goods and reasons for action. So although Williams's account and mine may turn out to be mutually reinforcing, they have their own separate tasks to perform.

However, the position defended in this first chapter may also help us to correct a misconception that I believe is common to Sidgwick, Rawls, Nagel, *and* Williams. All four philosophers appear to assume a dichotomy between, on the one hand, taking all the times of one's life as equal in status and as equally concerning one, and on the other, giving special importance to the present (or some different, indexically-specified time).[33] But this may be a false dichotomy if time preference can be established in the above manner; for, as I have already been at pains to point out, our previous arguments have attributed no special status to the moving present or to any other indexically-designated time of life.

V

Perhaps the main impetus behind the rejection of time preference has been the view that temporal equality is necessary to any proper conception of the unity of one's own (or any single) life. How then, finally, can the present discussion defend itself against the charge that, by giving unequal importance to different times or time-periods, it forces one to regard human lives as less unified than they actually are?

Consider, to begin with, the pure preference for what happens later in life that we defended above. Such preference cannot, I think, be criticized on the sole grounds of failing to treat earlier and later periods as parts of a single life, but perhaps the criticism will be that it fails to treat these periods as equally *real* (parts of a single life). In that case, however, the critic must hold that treating times as unequally important (to the goodness of lives) involves treating them as unequally real, and on the face of it, this assumption has little to recommend it. The analogue, in the case of individuals, would be to maintain that a hereditary aristocracy that treated some people as inherently more important, or higher, than others was committed to the lesser reality of the latter, and both these parallel assumptions seem uncomfortably similar to

[33] See Sidgwick, op. cit., pp. 381 f.; Rawls, op. cit., pp. 293 ff, 420 f.; Nagel, *The Possibility of Altruism,* pp. 60 f., 101; and Williams, op. cit., p. 209.

the ancient but presumably now discredited view that reality admits of degrees and is proportional to goodness or perfection.[34]

Indeed, something more positive can perhaps be said to defend pure time preference from the charge of failing to appreciate the equal reality of different times of life. Someone who understands the character of his own life must have some sort of view of its different periods, but must also be aware of its finitude. But this fact of finitude has important repercussions for our attitudes towards the different epochs of a single life. Older people sometimes envy the young for having so much of their lives left to live, and the young, in turn, often feel sorry for older people because they have so little time remaining. Having a substantial amount of time left is thus often thought to be of positive value, and judgements about how fortunate a given person is at a given time seem to depend not only on what is happening to him and what he is doing at that time, but on our estimation of how much time the person can reasonably count on in the future.[35] (There is thus something ironic about very old people who are still actively engaged in careers, for their undiminished powers seem to mock the second-order inability of those powers long to sustain themselves—compare the pseudo-rejuvenation of those about to die of starvation.) In that case, our preference for later happiness and success may in some way reflect a desire that the later parts of a life should receive some compensation for the fact that time is running out. The greater importance attributed to later periods may thus represent the idea of a larger overall equality that emerges when the different periods of life are seen against the background of non-existence and death. And when one sees pure time preference as a kind of

[34] I do not see how these assumptions are established by anything Nagel says in *The Possibility of Altruism*—see, especially, p. 79 n.

[35] Such temporal values are no more inherently suspect than the value we attribute, e.g., to have a certain surgical operation over and done with.

balancing of the goods contained in a life against the structural (dis)advantages of different times of life, the criticism that it disregards the equal reality of different times of life, or the unity of life, may simply fall away. If we lived forever, our (detached) judgements about what makes lives good would perhaps put no weight on the distinction between earlier and later periods. But the fact is that life is finite, and Nagel himself stresses that finitude as one of the main features of a conception of reasons for action whose principal virtue is said to be its adequacy to the unity of life, its ability to treat life as more than a 'series of episodes'.[36] I agree with Nagel that the unity of life must be seen against the background of non-existence in order to be properly understood; but what I have said just above suggests none the less that far from entailing Nagel's temporal egalitarianism, such an understanding of life's unity may actually be favourable to pure time preference.

However, we should also consider whether our defence of lukewarmness to the characteristic goals and interests of certain periods of life may not underestimate life's unity by allowing people to be invidiously dissociated from certain real parts of their lives. But how such dissociation can be invidious if it is also typical and widespread is not easy to understand. If the idea is supposed to be that life is more unified than people realize, than their attitudes imply, what happens to the ideal of justifying metaphysical and moral conceptions partly on the basis of normal human self-understanding, an idea Rawls, Nagel, *et al.*, themselves invoke? But perhaps the worry that in the present discussion, human life is somehow being disassembled before our eyes, can best be met by pointing out an important kind of unity that our ideas allow to individual lives. For I believe those ideas ascribe (or can be extended so as to ascribe) to human lives a unity analogous

[36] See Nagel, *Altruism*, pp. 38 f., 60, 73.

to the (satisfying) unity attributable to the development and decay of organisms.

Consider how ordinary people and biologists tend to think of plants and animals over time. Within the life cycle of a given organism a distinction is typically drawn between periods of development and periods of decay, and this distinction is partly marked by treating a certain period of maturity as representing the fullest development of the organism and other periods as leading 'up to', or 'down from', it. In keeping with these distinctions, there is also a tendency to think of organisms as being most fully what they are (what they have it in them to be) during maturity, a tendency perhaps most clearly exemplified in the tradition of making general reference to organisms by their adult names rather than by names appropriate to other stages of their life cycle.[37] (We speak of the parts of a tree's life, not of a seed's or sapling's life; of the development and decline, or old age, of a horse, but not of a colt.) These various habits of thought and nomenclature imply a bias in favour of the mature life of organisms that rather parallels what we have been saying here about the prime of human life.[38] But if such bias is compatible with, even characteristic of, the unity (unified identity over time) attributable to organisms, then our own emphasis on the differential importance of different life periods is compatible with a notion of life's unity modelled on the unity that organisms possess through time, and I can see no reason to believe that such a unity is anything less than what we should expect, or want, from human life.

On the other hand, a biologically-modelled conception of life's unity need not overestimate or oversimplify that

[37] Cf. J. H. Woodger, 'On Biological Transformations', in Le Gros Clark and P. Medawar, eds., *Growth and Form* (essays presented to D'Arcy Thompson), Oxford: 1945, p. 95; and David Wiggins, *Sameness and Substance*, Oxford: Blackwell, 1980, p. 64.

[38] In keeping with what is said about the mature period of organisms, we have a tendency to think that people must attain (forge) an identity rather than always having one. (The prime of life presumably *begins* when identity has been attained.) Cf. Wiggins, op. cit., p. 167; and Erikson, loc. cit.

unity (e.g.) by presupposing the 'Simple View' of identity discussed by Parfit and attributable to Butler, Chisholm, and others. For the biological model does not say that identity through time requires the existence of something that undergoes no alteration; and although someone holding the Simple View will (according to Parfit)[39] be impressed by the fact that 'all the parts of a person's life are as much parts of his life', anyone who claims, with us, that some parts of a person's life are more important than others, is surely *not* going to be *impressed* by such a fact—even if he in some sense accepts it.

I do not claim to have said very much here about the nature of the unity-through-time of humans or of organisms generally, only to have defended the compatibility of what I have said about human life with a plausible notion of the unity of such lives that needs working out in detail.[40] And given the prevalence in Aristotle's thought of the organismic model of identity through time, it is not in the end very surprising that one can find anticipations of our present defence of time preference, and of its higher estimation of the goals and judgements of mature individuals, in the *Nicomachean Ethics*.[41]

[39] Op. cit., p. 141.

[40] For the interesting beginnings of such theory about the unity of lives and the unity-through-time of organisms, see Wiggins, op. cit., *passim*. Wiggins relies heavily on Aristotelian ideas about identity over time.

[41] See Book I, Chs. 3 and 9, for example.

2

Relative Virtues

Chapter 1 argued, among other things, that we need the notion of a good, relative to a time of life, in order to do justice to our own attitudes towards human well-being. In the present chapter, I would like to continue this line of thinking and show, in particular, that certain neglected aspects of human virtue are best approached via the notion of a relative virtue. Although most discussions of virtue, or human excellence, assume an absolute distinction between virtues and non-virtues, a more fine-grained approach is sometimes called for; and I shall argue that various facts about human life and development make it plausible to regard certain personal traits as virtues or excellences only in relation to some particular period of life, rather than as virtues, or excellences, *tout court.*[1] Moreover, the reasons we have for relativizing certain virtues to times of life naturally extend to the idea that a virtue in some possible world may actually be a vice or defect in others, to the idea that some virtues are world-relative. (But we shall see that none of this counts against the objectivity of our thought about virtue—relative virtues need not entail virtue-relativism.)

The case for relative virtues is best introduced by considering some of the things philosophers have wanted to say about 'life plans'. Life plans are often thought to display, in a high form, the virtue of (practical) rationality, but by noting some limitations and problems with ideas about life plans recently advocated by John Rawls, Charles Fried, David Richards, and John Cooper,[2] I hope we may be

[1] I don't think it necessary for our purposes to distinguish among virtues, excellences (of character), and good traits (of character).

[2] Rawls in *A Theory of Justice*, Cambridge: Harvard, 1971, esp. pp. 398–449, 550–63; Fried in *An Anatomy of Values*, Cambridge: Harvard, 1970, pp. 1 f., 98–101, 156–75, 234; Richards in *A Theory of Reasons for Action*,

led to see how various facts about growth and maturity recommend the fine-grained conception of virtue I shall be defending.

I

In recent years, the concept of a life plan has come into (renewed) prominence as part of the conceptual apparatus of ethical theory. John Rawls, for example, has treated the notion as essential to the explication of the idea of individual good (of individual well-being over time)[3] and John Cooper assigns the notion of a central role in Aristotle's ethical thinking in an attempt to harmonize and render plausible the bewildering complexity of things Aristotle says about *eudaimonia.*[4] But more importantly for present purposes, Rawls, Cooper, Fried, and Richards all treat the having and following of a life plan as a reasonable and desirable way to live. The first three seem to regard life plans as governing the whole (remainder) of a life, whereas Richards seems to think of life plans as involving 'extended periods', though not necessarily of the whole of a person's future.[5] But they all agree that people should adopt life plans at some appropriate time in the midst of their lives.

Rawls, for example, treats it as reasonable that one should adopt a life plan in choosing a career, but he makes very little of the attendant presupposition that children lack life plans. And, indeed, with the exception of Richards, all the above-mentioned philosophers treat life-planfulness as a univocally desirable form of rationality and say nothing to rule out the possibility that it might be regrettable, even deplorable, that (most) children and adolescents lack life plans. But this

Oxford: Clarendon, 1971, pp. 29–46; and Cooper in *Reason and Human Good in Aristotle*, Cambridge: Harvard, 1975, pp. 94–127.

[3] Op. cit., pp. 407–24. [4] Loc. cit.

[5] In work subsequent to *A Theory of Justice*, Rawls places some emphasis on the revisability of life plans, but takes back none of his earlier views about life plans. See, for example, his 'Kantian Constructivism in Moral Theory', *Journal of Philosophy* 77, 1980, esp. pp. 525–9, 544–8.

precisely leaves out of account how unnatural and unfortunate it is or would be for very young people to have life plans. In what follows I hope to show that the too-early possession of a life plan is a positively undesirable thing, and that rational life-planfulness should consequently be thought of as a virtue only relative to certain times of life. But it will be easier to understand why, if we first consider some other limitations on life plans that Rawls, Cooper *et al.* leave largely unmentioned.

The four philosophers under discussion all agree that actual limitations on our knowledge make it inappropriate (impossible) for life plans to be very specific about the remote future. Richards alone, however, makes the further claim that it is sometimes better to do without a life plan altogether until certain sorts of information become available. Where one cannot anticipate one's future desires or abilities, for example, making decisions—even conditional decisions—in advance, may adversely affect how well things turn out. A forceful example of this not mentioned by Richards, but particularly relevant to present purposes, arises in connection with (married) couples who want to pursue distinct careers. An academic woman with a non-tenured position in the same town where her husband has tenure, may well worry about the possibility of not receiving tenure. Should she perhaps then plan for that contingency, deciding in advance whether, in the event of being denied tenure, she should look for employment elsewhere (and if necessary spend much of her time away from her husband) or should stay with her husband (even at the cost of remaining unemployed)? Of course, other relevant factors may require her plan(s) to be more complicated than I have suggested, but in addition to the issue of appropriate complexity, there is the further, important consideration that the very having of such a (complex) plan may actually impoverish the choice it eventually leads our woman to make.

If, for example, she decides in advance to live apart from

her husband if that is the only way she can pursue her career, may that decision not make her slightly withdraw from him during the period of uncertainty about her tenure, so that it becomes easier for her to follow her plan if and when she is denied tenure? By the same token, if she decides in advance to stay with her husband, may that not affect the seriousness or the energy with which she pursues her career during the period of uncertainty about tenure? If we assume that plans (are likely to) have emotional repercussions in this way, then there will be excellent reason for the woman *not* to make an advance decision about what to do if she doesn't receive tenure. For even if she eventually does receive tenure, the woman is likely to cramp either her career or her marriage by deciding what to do in advance, and if she does not make contingency plans, then although her eventual decision may be full of storm and stress, she will not at least have limited herself *unnecessarily* or *prematurely*.

I think many of us would agree—and for reasons of the sort just mentioned—that it would be better for the woman not to decide in advance. But we would also find it understandable that anxiety about tenure might make her rush into a decision about what to do in the case of non-tenure (as if anxious uncertainty over an event could somehow be allayed by certainty about its effects on one's actions). But I think we would agree that the woman would do better to contain, or learn to live with, her anxiety, rather than allow it to force an issue that is better delayed. Of course, if the decision *is* delayed, then both her career and her marriage may flourish in the interim and her decision between them become all the more difficult to make. Indeed, a certain awareness of that fact may be part of what impels the woman toward an early decision. But such a decision would none the less be premature: it is worth waiting for a (possible) more difficult later decision, if that difficulty would be largely due to the greater richness of the then available alternatives.

Now one of the chief attractions of the notion of a life

plan is that it seems to allow one to be active in determining one's fate; that it permits one (in a phrase both Rawls and Cooper use) to 'take charge of one's life'. On the whole, this is no doubt a rational and valid motive, but the above case suggests that it is not always so. Sheer suspense about whether she will be given tenure, sheer tension (and anger?) about having to be passive before the decision of others, might induce our woman to (re)assert some measure of active control over her life by at least having a plan about what to do in the event of not receiving tenure. But in her situation this is likely to be the wrong thing to do, and a sign, actually, of her weakness in the face of anxiety. Rawls tells us that one's desires and capacities are affected by which life plan one follows: what one does according to the plan eventually makes a difference to the kind of person one is. But in the case of the academic woman, it is clear that the adoption of a life plan for certain eventualities may have adverse effects even before those eventualities arise and before one follows what the plan dictates with respect to them, and the example thus suggests some significant limitations on what is appropriately the subject of life plans.

Let me now mention yet another respect in which too much has been included within the scope of life plans through insufficient appreciation of the importance of a certain kind of passivity, that of waiting rather than taking charge, in various areas of our lives.

For Rawls, Richards, Cooper, and Fried, the (basic) goods of human life properly figure as ends or goals (to be optimally catered to) within rational life plans or schemes of conduct. But to think of personal goods as ends figuring within life plans, is to regard them as things to which one can take means, as things one can try to achieve or obtain (or else as things under the immediate control of the will). And it seems to me, rather, that some of the most basic goods of life are precisely not things it makes sense to think of pursuing or controlling in this way. I have in mind here not merely assets, like a strong constitution or high intelligence, that one either has or lacks throughout one's life, but also certain goods that come, or do not come, during the *course* of a life: in particular, the goods of friendship and love. Because love, or the state of being in love, is generally recognized to be largely outside our rational control, people do not usually take steps or exert efforts to attain it. Yet we (nowadays) tend to think of love as a basic human good, as something *indispensable* to human flourishing in a way that certain other great goods (e.g., the enjoyment of beautiful music) may perhaps not be. And so it makes little sense in human terms for the good of love to feature among the ends catered to in life plans that dictate the means to, the best way of achieving, those ends.[6] The enthusiasts of life plans have stretched the notion beyond its actual usefulness, because some goods cannot sensibly be treated as goals of a plan of action.

Of course, love may sometimes require a certain amount of 'stage-setting' and if one really wants (hopes for) love to figure in one's life, one will avoid setting obstacles in its way, e.g., by signing up with the French Foreign Legion. But none of this need entail that we can reasonably plan to fall, or be, in love (or plan to fall, or be, in love if and when certain conditions obtain). Nor does it entail that we should take means to this good rather than waiting for it to occur and expecting, in the light of some knowledge of human nature, that in due course it probably will.[7] I do not, however, mean to suggest that one must be entirely passive during love's genesis. Clearly love can develop out of the activities of an ongoing relationship, but these activities are not, I

[6] Although many people hope and expect to get married someday, it would be odd for someone with no particular person in mind to say he planned to get married someday, and the reason, I assume, has to do with the fact that marriage is thought to presuppose love and that we cannot reasonably plan for love. (I am not, by the way, attributing the unplannability of love to its supposedly unselfish character. Life plans may certainly include unselfish ends of action.)

[7] Even getting psychoanalyzed in an effort to become more capable of love is more plausibly regarded as a clearing away of obstacles to falling (being) in love than as a means to (part of a plan for) falling (being) in love.

believe, typically directed towards creating the love that grows out of them. The fact that we do not pursue the condition of love means that we are in some sense passive *vis-à-vis* its coming; but it does not follow that we are passive in every relationship out of which (we hope) it may eventually grow.

All the above seems to be true not only of love, but of the (mutual) affection of friendship, which thus constitutes another exception to the life-plan theorist's assumption that the various goods of life figure as goals within the life plans of reasonable individuals. Rawls tells us that we are happy when we are in the way of successfully carrying out a rational life plan; but if important goods escape the net of such plans, then even if the lack of love or affection may not interfere with the fulfilment of what sensibly can occur within a life plan, it may still prevent the happiness of an individual who is doing all he has planned for (and could reasonably hope to plan for). Rawls, Richards, and Fried all explicitly claim that acts and relations of love and friendship are among the ends we pursue and are thus among the goods that must be balanced off or orchestrated within life plans. And they also hold that love and friendship involve the pursuit of certain ends, e.g., the well-being of the loved one or friend. But it is one thing to say that love *involves* (the having of) ends of action, quite another to claim that love is *itself* an end of action, and even if it makes sense to accept the first of these claims, the above discussion may help to explain why the second should be rejected.

II

We have just been exploring two related limitations on life plans. The idea that all goods are ends that a life plan can take into account ignores some crucial respects in which we must be patient rather than active about our own good. And the example of the academic woman illustrates a way in which our good depends on not asserting a premature planful control over our future. What I would like now to suggest is that it is similarly undesirable for a very young person to have *any sort of life plan at all*. Children are not ready to take active control of their lives, and when one of them does plan out his life in the sort of detail that is thought to be appropriate to older individuals, we may well suspect that he is acting from anxiety or impatience about the future and would be better off if he didn't feel the need for active long-range planning and simply let the future take care of itself. The academic woman of our earlier example may have to make a conscious decision not to plan for the contingency of non-tenure, but in the normal course of things, a child will not even *think* about having a life plan.[8] Rather than decide not to take charge of his life, he will just *go on* not taking charge of his life, not having a life plan, being cared for by his parents or guardians.

Consider, for example, a child at primary school who plans to go to a certain medical school and become a surgeon. We all know or have heard of such children; and their vast predetermination is likely to seem suspect—though somewhat awesome—to most of us. If the child's father went to that medical school and is a surgeon, we shall wonder whether the child doesn't feel some sort of more or less explicit pressure to be like his father which one hopes will eventually dissolve. And will we not think that his plan may well collapse when he goes away to college, or before, and, more to the point, that if it does not, that will mean that he is probably not allowing life to influence or change him, that he is too rigid? His planfulness will in that case prove positively detrimental to his development: it will prevent him from seeing, for example, whether he might not prefer pure science, or the law, to medicine, or prefer another branch of medicine, in the light of subsequently uncovered and developed interests and talents. And similar points could be made about his choice of medical school.

[8] How much of what follows also applies to (older) adolescents, I leave an open question.

Of course, some people are in a position to choose a career a good deal earlier than others are. But even so it seems to me that children are not ready to decide on a particular career. A child with remarkable talent for tennis, say, or for the violin may want to develop his skills to the utmost. But does that require him actually to decide on a career in the relevant area and make definite plans regarding its various stages? Wouldn't it be more desirable and more appropriate to his time of life for him to remain open to other possibilities?[9] A child does better to wait until his environment calls for definite choice of career; and this precisely means not prematurely attempting to take active control of certain aspects of his life, but rather being patient and trusting to the future.[10]

Thus rational life-planfulness is a virtue with a temporal aspect. It is, let us assume, a positively good thing, a virtue, in individuals mature enough, for example, to decide upon a career or profession. But in the very young this same trait is, as we have argued, undesirable, a sign of excessive pressure or anxiety and a brake on development. And this change from undesirability to desirability is perhaps best captured by the notion of something's being a virtue relative to a time, or period, of life. Once one has reached a certain point of maturity, life-planfulness is a virtue of practical rationality; but the disposition to have and follow a life plan is the opposite of a virtue, an anti-virtue, in (relation to) childhood.[11] (The term 'vice' has unfortunate connotations

[9] All this is consistent with allowing that those who make premature life plans concerning careers are sometimes very successful in those careers. But such premature choices are typically the result of parental pressure, and those who yield to, and succeed under, such pressure can hardly help being emotionally scarred by it as well.

[10] Some children have to live by their wits and become 'street wise': but this doesn't involve rational life-planfulness in the sense intended by Rawls *et al.* Similarly, a child who simply wants to be a fireman or President someday lacks a life plan of the sort we are speaking of. Of course, some societies, and some parts of our own society, demand an early choice of occupation. But what is at stake in such cases is typically not a *career*, and in order to simplify the discussion, I shall follow Rawls *et al.*, by concentrating mainly on the choice of careers.

[11] For doubts about whether life-planfulness is a virtue even in adulthood, see, e.g., Bernard Williams, 'Moral Luck', *Proceedings of the Aristotelian*

which somehow make it inappropriate for the general designation of qualities opposed to virtue. We do not, for example, call impatience a vice, even though it is the opposite of a virtue, so perhaps the barbarism 'anti-virtue' is the best term for describing undesirable character traits like impatience and premature life-planfulness.)

Much of what has just been said about life-planfulness carries over, furthermore, to another, more familiar virtue of practical rationality, the virtue of prudence. Prudence—in the ordinary English sense, rather than as *prudentia*, or practical wisdom—is clearly different from rational life-planfulness: it entails a tendency to caution, to playing things safe, that need not come into every life plan and that can perhaps exist in the absence of any definite life plan. (The life plans of a religious crusader might allow little scope for caution, prudence, and the like; and an individual who prudently planned to avoid certain risks and catastrophes might not have any plan for achieving his positive ends or a very definite view, even, of what he wanted from life.) But despite these differences, the virtue of prudence displays the same temporal features that make it natural to think of life-planfulness as a period-relative virtue. Prudence involves a commitment to long-range planning—a child who exercises reasonable caution in crossing streets does not thereby qualify as prudent in the ordinary sense. And prudence thus understood seems to be a virtue in mature individuals.[12] But it can also be displayed prematurely, and when it is, prudence is the very opposite of a virtue.

Imagine, for example, a schoolboy who takes out an insurance policy on his own life in order to protect the

Society, supplementary volume, 1976, pp. 115–35. Incidentally, our discussion of the limitations of life-planfulness is not supposed to undercut Rawls's use of the notion in justifying his conception of justice.

[12] Some of the above points about prudence can be found in J. D. Mabbott and J. N. Horsburgh's contributions to a symposium on 'Prudence', *Proceedings of the Aristotelian Society*, supplementary volume, 1962, pp. 51–76. In fact, Horsburgh tentatively suggests that prudence may be undesirable in the young and not *very* desirable in older individuals.

interests of his future wife (whoever if ever she may be) or one who saves half his allowance against the needs of his old age. I believe cases of this sort actually sometimes exist among the very young, but I cannot think of any example of childhood prudence that does not immediately seem odd, inappropriate, even pathological. Very young people have no business being prudent, and when they are, they demonstrate an unfortunate lack of trust in the world around them. When all is going well, a child simply and unreflectingly trusts to the future and his parents or guardians exercise any prudence that needs to be exercised in his behalf.

We have seen that those who lack life plans and take things as they come also trust to the future, so by virtue of their common emphasis on the long term, both prudence and life-planfulness can be said to stand opposed to a certain attitude of trust that is naturally, though not always, found among the very young. But such trustingness is at best only a desirable characterstic in the young. Older people who trust to things the way a child does, are spoken of, disparagingly, as 'trusting souls', so this same trait seems actually to be an anti-virtue in relation to adulthood. Trustingness may thus count as a childhood-relative virtue, and it is worth noting that neither the sheer 'passivity' of this character trait nor the fact that a trusting individual may have no contrary temptations creates an automatic barrier to treating trustingness as a virtue. There is something equally passive about the traditional virtue of innocence, and, as Philippa Foot has pointed out, an individual's perfect honesty may precisely *consist* in his freedom from certain temptations.[13]

On the other hand, in order to show that life-planfulness and prudence are relative virtues, it is not absolutely

[13] See the title essay of *Virtues and Vices*, Oxford: Blackwell, 1978, pp. 10 ff. Of course, prudence is often spoken of as a virtue, whereas trustingness almost never is. But I think this only shows that we have an unreflective tendency to concentrate on the period of adulthood in considering the desirability of various traits. There is nothing wrong with saying prudence is a virtue if one is prepared to admit that one is limiting one's focus to adulthood and speaking elliptically. And have we any reason to resist such an admission?

necessary to show that trustingness is positively desirable in, and a virtue relative to, childhood. It is enough that the childhood anti-virtues of prudence and life-planfulness become virtues in (are virtues relative to) adulthood, and the fundamental idea of virtue-relativity is perhaps best exemplified in these particular transpositions of virtues and anti-virtues.[14]

But if the notion of time-relative virtue is needed to give the full complex picture of the sort of goods prudence and planfulness are, it is also needed to account for our attitudes towards other supposed virtues. I mentioned earlier that innocence has traditionally been regarded as a virtue. Today, however, I think a large number of people would disagree with that assessment, and although I have no intention of trying to settle this disagreement, the notion of virtue-relativity may help to clarify what is at issue here and provide a middle ground between admirers and disparagers of innocence.

Opponents of a virtue of innocence often refer to the childlikeness of adults who display that quality, so perhaps their criticism of innocence need not involve a total rejection of the virtue, but only its relegation to the status of a childhood-virtue. Perhaps one should find innocence absurd, pathetic, or gratuitous in an adult, but desirable, admirable, or possibly even enviable in children, and such an attitude finds its natural expression in the thought that innocence is a virtue in (relation to) childhood, but the opposite of a virtue in (relation to) adulthood. Some will undoubtedly wish to reject this compromise, holding either that innocence is a virtue, and perhaps especially precious, in adulthood, or else that there is nothing particularly desirable even about childhood innocence. But the options are in any case more complex than may originally have been suspected,

[14] The childhood-relative personal goods of Ch. 1 do not similarly transpose into adult evils, or anti-goods. Unfortunately, I have no ready explanation of this divergence between the relativity of goods and the relativity of virtues.

and their full articulation requires us to go beyond absolute conceptions of virtue. The idea that innocence is (merely) a childhood virtue may not be obviously correct, but it is a possible attitude, and far from an absurd one, and it cannot, I believe, be expressed without implicit recourse to the notion of a time-relative virtue.

III

It is worth noting that certain virtues or excellences—among them the traditional cardinal virtues of courage, wisdom, temperance, and justice—lack some of the features that make it appropriate to speak of prudence, life-planfulness, and (possibly) innocence as virtues (or excellences) only relative to certain times of life. Prudence is only a period-relative virtue because there are times of life—most notably childhood and adolescence—when circumstances can produce prudence in an individual, but the trait is positively harmful or undesirable, an indication of something wrong with the individual and/or his environment. But a minor virtue like patience seems both to be capable of existing and to be desirable in every age group, even (perhaps especially) among children, and there is no reason therefore to suppose that it is anything but a virtue or excellence *tout court*. It is similarly difficult to think of a time of life when one of the four cardinal virtues could occur and yet be undesirable or harmful (to its possessor). (A wise or just child—like the twelve-year-old Jesus lost among the scholars in the Temple—might initially embarrass his parents; but would such virtue, given favourable circumstances, be harmful or useless in some way that adult wisdom or justice is not?) Isn't the problem with the cardinal virtues really the very opposite of this, namely, that it is all too easy to think of times of life when they are desirable and even needed, but rarely, if ever, occur?

It is proverbial—but not for that reason implausible—that wisdom comes to individuals, if at all, only late(r) in life.

And if that assumption is correct, then this cardinal virtue is not as usefully keyed to the human life cycle as such minor and relative virtues as prudence. Prudence may at some times be undesirable, but it has a tendency to exist precisely when it can do the most good—prudent children and adolescents being rarer than prudent adults. (Practical) wisdom, on the other hand, includes a sense of the relative importance of things, and would surely be of use to a young adult attempting to formulate some sort of acceptable life plan. But we are assuming that wisdom is never present so early in life, so with respect at least to this one virtue, there seems to be a lack of fit between human needs and actual human powers: wisdom seems to be available only long after the time at which it could do us the most good, only at a vantage-point later in life from which a person can look back and see how wisdom might have enabled him to avoid the very mistakes that helped to make him wise. (As we say: if youth only knew, if old age only could.) There is something ironic, perhaps tragic, in this fact of bad timing. And yet advocates of rational life-planning, whose ideas make this irony assume its starkest form, never mention the divergence between the good, wisdom ideally could do and the good it actually accomplishes in human life. And if Aristotle counts as a life-plans theorist, then the omission is especially odd in his case, since he seems explicitly to deny that young men and women are ever wise.[15]

IV

We have seen that certain traits of character may be virtues (only) relative to certain times of (human) life. But if what is virtuous can vary over time, then there seems to be an element of contingency in what counts as a virtue, and we may perhaps wonder whether any deeper form of contingency infects familiar forms of virtue and whether, in

[15] See the *Nicomachean Ethics*, Book I, Ch. 3; Book VI, Ch. 2.

particular, some virtues might not have counted as virtues at all (time-relative or otherwise) if the world, and we, had been different in certain ways.

Such contingency has, in fact, been defended by Philippa Foot. In her paper 'Virtues and Vices', she claims that virtues are in some sense corrective, so that if, e.g., there had been no human tendency to give in to certain temptations, temperance would not have been a virtue, and if mankind had been subject to weaknesses different from those that actually beset us, there would have been virtues that do not count as such given the actual state of things. This view introduces an element of world-relativity into our conception of familiar virtues, but it is worth noting that it falls short of the radical contingency that would be suggested by a full analogy with time-relative virtues. Foot never claims that temperance would count as an anti-virtue, an undesirable trait, if things were different enough, only that it might fail to be the good or valuable trait of character it actually is.[16]

However, the example of time-relativity might encourage us to ask whether it may not be possible for certain traits to count as virtues in some worlds (or relative to certain assumptions about the world), but represent positively undesirable traits in other worlds. And a suggestion to this effect has indeed already been put forward by P. H. Nowell-Smith, who claims that malevolence would be a virtue and benevolence a vice if people were so stupid and inefficient that attempts to do good (always or typically) resulted in more harm than good.[17] I think we should have our doubts about whether this description really represents a possible state of affairs, but even so, there may be other qualities whose status as virtues or anti-virtues can more easily be thought to depend on basic facts (assumptions) about the

[16] Given her discussion of what it is for something to be a virtue in one person but not another, there may be a way for Foot to argue that temperance could (conceivably) have been an anti-virtue. But I prefer to explore a different and to my mind more interesting way in which virtues may be world-relative.

[17] In his *Ethics*, Harmondsworth: Penguin, 1959, p. 250.

world. Even if it is difficult to imagine in any realistic detail what it would be like for malevolence to be a virtue, I think certain traits of rationality may perhaps plausibly be said to be virtues under certain assumptions about the world, but undesirable, anti-virtues, relative to others, and I would like to explore this possibility in what follows. We have previously seen that certain sorts of practical rationality may be desirable or undesirable depending on one's time of life, and I believe that whether a certain kind of *intellectual* rationality counts as a virtue or anti-virtue may in somewhat similar fashion depend on certain general assumptions about the world. Let me explain.

A fairly typical anti-religious rationalist will want to claim that the hypothesis of (a Judaeo-Christian) God cannot be reasonably believed, but also cannot be absolutely excluded, on the basis of available evidence. But (he may continue) if religious believers are in fact right about God, then they are merely lucky, for their irrational credulity, their faith, is a defect of intellect even if it actually leads to truths or good results. The rationalist could then claim that his own unwillingness to go beyond the evidence with respect to the important issue of God counts as an intellectual virtue quite independently of whether there actually is a God. (For an example of such rationalism, see W. K. Clifford's 'The Ethics of Belief'.[18])

On the other hand, a religious fideist might claim that strict evidential rationality is a spiritual defect even if there is no God and that his own openness to faith should be regarded as a virtue 'come what may'. (Religious thinkers are usually unwilling to consider what would follow from God's non-existence, so it is difficult to find examples of the fideistic view that faith is, or would be, a virtue even if God doesn't, or didn't, exist. But one plausible candidate is William James in '*The Will to Believe*'.)[19]

But need we in fact adopt either of these antithetical

[18] In his *Lectures and Essays*, London: Macmillan, 1879, vol. II, pp. 182 ff.
[19] In *'The Will to Believe' and Other Essays*, Harvard, 1979, pp. 13–33.

attitudes? If despite what the sceptic takes to be a lack of evidence, there really is a God who desires and rewards the attitude of faith, then perhaps faith is something mankind needs for its own ultimate well-being, and the insistence on rationality in regard to the evidence for a deity an obstacle to individual salvation and to possession of the ultimate truth about things. So perhaps we should hold that if God doesn't exist, an unwillingness to believe in God on faith is a virtue, but that if certain religious assumptions are (were) true, then faith is (would be) a desirable and needed trait of human character, a virtue, and refusal to make the leap of faith an anti-virtue.[20] In that case, the virtuousness or anti-virtuousness of faith or strict rationality would depend on contingent (debatable) facts about man's place in the universe, and this clearly entails a strong form of world-relativity according to which certain virtues and anti-virtues transpose their status across possible worlds (or possible assumptions about the actual world[21]).

[20] I assume that faith need not involve, and in the present context should be thought of as *not* involving, any tendency to go beyond the evidence in extra-religious matters. I also leave it open whether in some possible world-situations, one might not need evidence in order to have a reasonable belief in some sort of deity. In 'Is Belief in God Properly Basic?', (*Noûs* XV, 1981, pp. 41–51), Alvin Plantinga has recently argued that certain religious beliefs might be as non-inferentially reasonable as normal perceptual beliefs if (there were a) God (who) built into human beings a disposition automatically to take certain phenomena as due to or coming from him. Beliefs acquired in this way would, he thinks, be as grounded (in the way things are) as any beliefs immediately acquired through perception. Since, however, I have by 'faith' been meaning an essentially irrational trait, the conditions I have in mind as allowing faith to be a virtue are not ones in which God makes belief in him non-inferentially rational along the lines Plantinga mentions. Rather, they are circumstances, worlds, in which belief in God is not (even other things being equal) automatic and in which God requires us to 'meet him halfway' in some sort of act of faith. For an example of theology of such a world, see John Hick's *Evil and the God of Love*, London: Macmillan, 1966. (For simplicity I shall not consider how wager-type arguments bear on all these issues.)

[21] The qualification is necessary because of the possibility that certain religious or anti-religious assumptions may be metaphysically incoherent. For example, the assumption that God is a certain sort of unembodied being might turn out to be necessarily false, and defenders of certain traditional arguments for God would want to say that it is metaphysically impossible for a world like ours not to have been brought about by a deity. In either case, we would have to speak of certain traits as being virtues relative to certain basic assumptions about the world, rather than as relative to certain possible worlds.

But there is an objection, a fairly obvious objection, to this way of seeing things, and in germ it is already evident in the attitude attributed earlier to the rational anti-religionist. The latter believes that if there is a God, religious people are merely lucky to have hit upon the truth in this matter, so that his own refusal to go beyond the evidence still counts as intellectually superior to the credulity of faith. But since a world-relative conception of the virtue of strict rationality makes the validity of his attitude hinge upon whether religious people are or are not lucky in their beliefs, it turns the virtuousness of his own strict rationality into a matter of luck. He will have a counterpart in some possible world who is in the same epistemological position as he is *vis-à-vis* the existence of God; for one of them the refusal to believe will be a virtue as for the other it is a vice; and it will be a matter of luck if he is the one for whom it is a virtue. And of course there seems to be something inherently objectionable in any connection between virtue and luck.

None the less, common moral thinking is deeply committed to such a connection. Thomas Nagel has pointed out, for example, that our moral assessment of an act done under uncertainty can vary from positive to negative, depending on the luck of circumstances,[22] and though he gives no specific examples, the following perhaps will do: a sentry given reasonable orders to kill strangers on sight may be unable to bring himself to kill an approaching pregnant woman and the toddler she has in tow. If she is as harmless as she appears, then he may be (thought) commendable for displaying a meritorious flexibility in interpreting orders. But if, on the contrary, she turns out to be a spy, we are likely to find no virtue in his flexibility and to think him condemnable for not paying scrupulous attention to duty. Thus on a quite ordinary moral view of things, the virtue or demerit of his act depends on features of the situation not

[22] In his contribution to the symposium 'Moral Luck', *Aristotelian Society*, suppl. vol., 1976, pp. 137–51.

subject to his control or certain knowledge and is therefore in some degree a matter of luck.

Of course, we may recoil from this way of seeing things, because of our tendency to connect virtue (vice) with praise (blame) and to feel that praiseworthiness and blameworthiness cannot depend on such extrinsic factors. But (as Nagel himself emphasizes) our tendency to assign moral merit and demerit on the basis of extrinsic factors runs very deep and is not easily renounced—perhaps, as Nagel also suggests, what one should, and can, give up is the association of virtue and anti-virtue with praiseworthiness and blameworthiness. There may be difficulties in the idea that the moral character of an act in some measure depends on luck, but we may have to live with them if the alternative is some (even more implausible) wholesale rejection of common-sense morality. And if moral luck is thinkable, possibly even acceptable, in regard to the character of particular acts, then perhaps the status of certain virtues as virtues can depend on a kind of *cosmic* (moral) luck. Thus even if we cannot *disprove* the rationalist belief that resistance to faith is a virtue come what may, we need the concept of a world-relative virtue in order to clarify what is *at issue* here. And indeed, only if we are prepared to think in terms of world-relative virtues, can we even see that there is a possible alternative to the traditional theistic belief that faith is an (absolute) virtue and the traditional rationalist claim that it is an (absolute) intellectual defect.

It may be wondered, on the other hand, whether the status of faith as a virtue can be made to depend solely on facts of human psychology, rather than on basic questions about man's place in the universe. If, given God's *non*-existence, man needs religious faith in order to be happy, or avoid being miserable, can we say that faith is a virtue, but would not have been one if faith had not been so necessary to his happiness? Perhaps we could, but it is interesting that thinkers like Voltaire and Archbishop Tillotson who have recommended faith as a salve to men's minds have also said that if God didn't exist, it would be better for the *generality of mankind* not to know it.[23] Such talk implies that in the relevant circumstances, the noble strong few ought to face the truth about God's non-existence, and the high estimation of truth implicit in this judgement seems incompatible with claiming that the tendency to have a false but consoling belief in God would be a virtue. In fact, the value we place on truth may help to make it clearer how faith can be a virtue in a world where (a hidden) God requires faith; for in such circumstances religious faith not only is necessary for man's salvation and well-being, but has actually got hold of the fundamental truth about things. Where faith is merely necessary to human happiness, faith may only be a (world-relative) personal good, but where it puts man into deeper touch with reality, it may also count as a (world-relative) virtue.[24]

If what I have been suggesting, then, is correct, an accurate account of human virtue cannot limit itself to speaking of absolute virtues and vices. Moreover, some of our most promising examples of relative virtues have been connected in one way or another with the notion of rationality. Earlier on we saw that some forms of practical rationality are desirable only at certain times of human life, and the present section has suggested that the virtuous character of certain sorts of intellectual (or evidential) rationality is plausibly thought to be relative to certain assumptions about the world. If the world is as the atheist assumes, then the rationality (or rationalism) that leads him to think of it that way is a desirable mental trait, a virtue, and the faith of the religious a mental defect or weakness. But just the reverse may be

[23] See Voltaire's Letter to Charles Augustin Feriol comte d'Argental, in T. Besterman, ed., *Voltaire's Correspondence*, Geneva: Institut et Musée Voltaire, 1962, vol. LXXI, pp. 246 ff.; and Tillotson's *The Works of Dr. John Tillotson*, Dublin, 9th edition, 1726, Sermon I, 'The Wisdom of Being Religious'.

[24] On the impossibility of treating what is directed towards falsehood as a virtue, see Thomas Aquinas, *Summa Theologiae*, Blackfriars: 1974, vol. 31 (Faith), pp. 15 f., 135. But Aquinas never considers the possibility of faith in God being mistaken.

true, if the theist turns out to be correct, and if it is his very tendency to go beyond (or against) the evidence (in this one area), it is his desire, in James's felicitous phrase, to 'meet God half way', that puts the theist in the way of truth and salvation.

Now in advocating these views, I have tried to stay clear of anything that entails moral relativism. Roughly speaking, relativism denies the objectivity of value judgements and does so via the claim that their validity or appropriateness is a function of what people believe or choose. But the idea that certain virtues are virtues only in relation to certain times of life or certain general world-situations makes the status of such virtues depend on facts about how human needs vary over time or on deep (contingent) facts about man's place in the universe, rather than on anything subject to human choice or affectable by human belief. So a belief in relative virtues need not force us to any sort of relativism about the virtues, and implies only that any correct or objective account of these matters requires some fine distinctions not usually to be found in discussions of human virtue.

3

Dependent Goods, Dependent Virtues, and the Primacy of Justice

Discussions of virtue and human good generally ignore relative virtues and personal goods; but they also typically neglect the fact that many virtues only count as such when they are attended by certain other virtues and that some of the things most sought after in life are (considered) good only to the extent that certain further goods underlie them. Some goods and virtues, in other words, depend (respectively) on other goods and virtues for their value, and in this chapter I shall offer an account of this notion and attempt to show how pervasive it is in our understanding of moral phenomena, and how necessary to any complete description of what we value in individuals, relationships, and whole societies. In particular, I shall argue that the notion of dependency can be used to add an important dimension to the Rawlsian view of justice as the first virtue of social institutions. But let me begin with some illustrations.

I

Consider humility as a secular (non-theological) virtue. There are certain people whose personal qualities we do not much admire, but who do not puff themselves up about non-existent virtues. Most of us would not particularly esteem such people for their humility, and the reason, I believe, is our conviction that they have, as one might put it, *so much to be humble about*. Such people lack, of course, the *vice* of overweening pride, but it is only in people with desirable or admirable qualities that humility shines forth as a (secular) virtue and represents an additional virtue or good trait beyond those we already recognize in a given individual.[1]

Now I have no ready account of what humility as a purely secular virtue consists in. For the present purposes let us simply assume that it is not some illusory form of self-deception in which one somehow contrives to believe oneself worse than one knows one is, but rather some desirable and positively characterizable way of avoiding vanity (that does not require religious submission or a 'broken spirit' and that may even be compatible with certain kinds of pride). But however we are ultimately to analyse humility, it follows from the little we have already said that humility is (in the sense intended) a dependent virtue. For intuitively it seems to attain its full status as a virtue or desirable trait of character only when accompanied by other desirable traits. It is a positive virtue only in someone we have other reasons to think well of. In addition, humility can seem more wonderful, more admirable, the more highly we regard someone's other traits; but although such quantitative covariation may also characterize the relation of virtue-dependence, I shall simplify matters by treating dependence as an all-or-nothing matter.

The status of humility as a dependent virtue has, indeed, some interesting repercussions. Philippa Foot has recently suggested that traits like industry and prudence can be virtues in some people, but not in others.[2] These personal qualities can, she thinks, be overdone, and they do not count as (act as) virtues in those who in fact overdo them. Our discussion of humility, however, suggests another way in which a trait may be a virtue in one individual and not another: namely, by being a quality whose status as a virtue in a given person depends on whether that person has various other virtues.

[1] We can also admire people for humility in the face of great wealth, beauty, or social status, but this does not substantially affect the main point, and I am simplifying.

[2] In the title essay of her *Virtues and Vices*, Oxford: Blackwell, 1978.

And humility is not the only virtue that can be characterized in this way. Unless we are prepared to hold—as most people would be reluctant to do—that anyone who acts from conscience automatically does what is right (for him to do), we are not likely to admire the conscientiousness of a Nazi prison-camp guard whose sense of duty dictated that he should simply follow orders, however 'disagreeable'. In the absence of common human decency, of a humane sense of values, conscientiousness seems utterly lacking in value, but we need not for that reason cast conscientiousness into the limbo of utterly neutral character traits. Rather, we can regard it as a dependent virtue, and I think this description conforms fairly well to the attitude most people (now) actually have towards conscientiousness. In someone with (the virtue of) common decency, conscientiousness may plausibly be thought to represent a further good trait, an additional ground for good opinion or admiration; but in the absence of basic decency or humanity, its presence is absurd or, in the worst cases, pernicious.

However, it is worth noting the different forms that dependency takes in the two cases of humility and conscientiousness. There are a broad range of valuable traits (and possessions) with regard to which humility seems praiseworthy, so although humility seems lacking in value unless it is 'about' valued traits, the range of valuable traits that are appropriate 'targets' of valued humility seems wide and open-ended. The status of humility as a virtue thus does not depend on the possession of any particular other virtue or good trait, and we may therefore say that humility is an unspecifically dependent virtue. Conscientiousness, on the other hand, seems to depend upon the particular, though by no means narrow, trait of decency (or humaneness). For although it can *exist* independently of the particular virtue of decency, conscientiousness possesses its status as a virtue, as a good trait of character, only where decency also exists (as a virtue). So we can call conscientiousness a *specifically dependent* virtue. However, in speaking as I just have of unspecifically and specifically dependent virtues, I have deliberately excluded consideration of another possible notion of dependency according to which one good characteristic (or thing) depends on another *for its existence.* Saintliness may depend on decency, for example, in the sense that the former 'virtue' cannot even exist apart from decency. But in the present discussion I shall only be interested in cases where a given trait—e.g., conscientiousness—can *exist* apart from another —e.g., decency—and only its *value*, its status as a virtue, depends upon the presence of the other trait (as valuable[3]). And in speaking of dependence—whether of virtues or of personal goods—I shall always have this latter sort of relation in mind.

We have just been discussing two rather commonsensical examples of the idea of dependency, and although the notion has, as I have said, been largely neglected by philosophers, it is worth pointing out that Kant, at least, makes fairly explicit use of it. In Chapter II of the 'Dialectic' of the *Critique of Practical Reason*, for example, Kant says that happiness is not absolutely good, but presupposes conduct in accordance with the moral law, or virtue, as the *condition* of its goodness. And in Book I, Chapter I, of *The Groundwork of the Metaphysics of Morals*, he speaks of the good will as the unconditionally good condition of every other good; and seems to want to say, in particular, that such traditional virtues as temperance and self-command are only good when possessed by someone with a good will.

I have, however, been unable to find anything similarly explicit elsewhere in the literature of ethics, so the present discussion can perhaps be viewed as an attempt to (re)awaken interest in the notion of dependence by setting alongside Kant's somewhat idiosyncratic opinions a picture of the dependence, and independence, of virtues (and, shortly, of personal goods) that better suits our implicit—though not,

[3] The words in parentheses leave open the theoretical possibility that a trait like decency may itself depend on other virtues. It is only where decency exists *as a virtue* that conscientiousness possesses its value.

for that reason, necessarily uninformed or untutored—views of these matters. But not merely that. The significance of the idea of dependence is to some degree also highlighted by the fact that we need to invoke this notion in clarifying historically important ethical opinions that may be alien to our own. Thus we need not *agree* with St. Paul in I Corinthians when he says 'if I have all faith . . . but have not love, I am nothing',[4] but we still need the notion of a dependent virtue in order to give a plausible account of what he is actually *saying*. And in fact once the concept of dependency is brought clearly into view, a wide variety of ethical phenomena can be seen to be related in a way that might otherwise escape notice. Such traits of individual character as humility and conscientiousness represent plausible instances of dependent virtues, but, as we shall now see, the idea of dependency also has application in the area of human relationships.

Many philosophers have held that a person can be harmed by deceptions or betrayals he knows nothing about and that, in particular, someone ignorant of his wife's infidelities is worse off simply in virtue of those infidelities, and quite independently of how good his marriage seems to him. Now without either accepting or rejecting this notion, consider what happens when we complicate the usual picture of unsuspected infidelity by imagining that both husband and wife are successfully deceiving each other and that, in addition, each has an attitude of trust towards the other despite his or her own deceitfulness. Where there is mutual fidelity, mutual trust is a (further) virtue of a marriage, but in the present case mutual trust seems to be out of place, rather than a particularly desirable or admirable feature of the relationship. And the natural inference is that trust is a specifically dependent virtue of human relationships that is neither valuable nor praiseworthy except when underlaid by basic fidelity, or trustworthiness.[5]

[4] I Corinthians 13.

[5] I am not assuming that 'open marriages' involve deceit or infidelity, and am simply restricting myself to more traditional types of relationship.

However, there is an obvious conceptual distinction between virtues, on the one hand, and the sorts of things that make individual lives good or fortunate, on the other; and although virtues (of individuals) must be regarded as good traits of character, even as qualities that it is good for an agent to possess, one need not regard them as qualities beneficial to those who possess them. In saying that something is a good quality for an agent to possess, we may simply mean that others are likely to benefit from his possession of that trait, so unless the Platonic thesis that virtue must benefit its possessor is somehow correct, the sorts of individual and relationship-characterizing virtues discussed just above may not figure among the personal life-goods of the individuals whose lives were there in question. However, there is no need to restrict our discussion of dependency to dependent virtues, for relations of dependence can also be found among the things that contribute to the goodness of human lives.

Some of life's goods, for example, derive from one's relations with other people. Just as a person who values fame rightly or wrongly considers himself to benefit from being the object of certain attitudes (and actions) on the part of other people, so too, and perhaps more plausibly, can we reasonably want, and think of ourselves as better off for having, the affection and respect of someone we ourselves love or respect. Consider, then, a particular person who is partially disappointed in these desires. Her career has recently taken a disastrous turn for the worse, and in the course of receiving a great deal of sympathy concerning what has happened from a close friend, the woman somehow discovers that her friend's previous confidence in her ability has been undermined by his knowledge of her set-backs. He thinks that she is probably not as talented, determined, or psychologically fit as he had previously imagined, and, without intending it, allows this attitude to seep out in the course of his attempts to console her about her misfortunes.

Whereupon, she somewhat angrily says to him: 'I want

your respect, not your pity'. [She *might* say: 'I don't want your sympathy, I want your respect', but it would be natural to understand such a statement as containing a tacit 'merely' or 'mere' in the manner of 'I don't (merely) believe it, I *know* it.'] The woman speaks as she does because, although she wants sympathy and needs affection at this moment in her life, she also feels personally demoralized by what has happened and more than anything feels the need to shore up her own confidence and self-respect. So her friend's sympathy is of no help to her if it represents mere pity and is not accompanied by—underlaid by—respect. In her state of depression or self-doubt, an attitude identifiable as pity simply fills her with further doubts and makes her situation worse. And the only sympathy that can really benefit her at this point is a sympathy which, because it accommodated a sense of respect for her abilities, might help her to regain the self-respect she cannot now muster. Of course, the respect of some disinterested party might also help her to recover from her current set-back, but the respect of a sympathetic interested party is what (we may assume) she feels the greatest need for. And the attitude of her friend denies her this. Thus in the circumstances, (experiencing) her friend's sympathy is a specifically (and unilaterally) dependent (possible) good that has value for her only if accompanied by what she can regard as respect.

Consider the next possible views that one may take of the relation between love and sexuality. In addition to the 'puritan' attitude that sex has very little value by comparison with love, and the hedonistic idea that sex has value independently of (the presence of) love, there is another possible view according to which sex is one of life's great goods, but possesses this value only in the context of a love relationship. And one does not have to accept this point of view in order to see that it occupies an intermediate place between hedonism and puritanism and represents a rather widespread attitude towards sex. Yet, once again, we cannot even begin to say what this attitude amounts to without invoking the notion of a dependent personal good.

II

Some of the most interesting applications of the notion of dependency occur, however, in connection with whole societies, or social groups. Civility and community, for example, are among the social virtues that figure in traditional conceptions of the good, or ideal, society. And what I would like now to argue is that both these virtues unilaterally depend upon another social excellence, the virtue of justice.

A just society characterized by the absence of any sense of community or by an absence of agreeable and graceful forms of personal interaction would surely lack something significant and valuable, but that lack would not, I think, deprive the fundamental justice of that society of its social importance and value. But consider the opposite case where a fundamentally unjust social order is marked, for example, by high civility. Imagine an antebellum South of graceful manners cinematically smoothed of all defiance and sulking on the part of the 'lower orders'. I have no doubt that the predictable graciousness of social forms and customs would have been agreeable to the well-off members of and visitors to that society, pleasant by comparison, for example, with what is to be found in modern cities and increasingly everywhere. But it was presumably easier to be charmed by the good manners, if one contrived to ignore or forget about the injustice of the social system in which these delicate forms of civility were flourishing. Someone aware of the moral evils of the system might still have found it pleasant to encounter its gracious forms and manners, but he would have been likely also to have a sense of irony about them. For against such a background of injustice, such gentility, such high civility, would naturally lose the appearance of a great social excellence and seem superficial and empty.

It is difficult to regard gracious manners as a particular merit of a given society if one recognizes the basic injustice of the system in which they flourish.[6] And if the status of civility (gentility) as a social virtue is largely undermined by (underlying) social injustice, then it would seem that the social virtue of civility specifically, and unilaterally, depends on the further virtue of (basic) social justice.

Something similar can be said about the social virtue of community. Consider the sense of community, the mutual affection, that (in the interest of illustrating an ideal type) we might attribute to the antebellum South: whites feeling affection for their black mammies, and for the darkies generally; blacks knowing their place and regarding with affection and awe the whites who took such benevolent care of them. The fundamental injustice of the whole social system would, I believe, once again eclipse that society's characteristic sense of community, largely devaluing it as a social virtue, as an admirable feature of the system. And in that case we have another example of a social virtue that comes into its own as such only against a background of social justice.

It is not difficult to acknowledge or recognize such dependencies of social excellence. But it is another thing entirely to *explain why* the absence of justice undermines the worth of civility and community; and on this subject I shall confine myself to some brief conjectures.

A Marxist, for example, might claim that such devaluation becomes understandable when one recognizes the civility and community of unjust societies as forms of ideology that

[6] I thus distinguish between the merits of a given society as a society and what benefits particular individuals. But even if given individuals may benefit from the civility of an unjust society, someone sufficiently aware of and disgusted by the injustice might take no pleasure in the surrounding gentility or find such pleasure, when it occurred, hollow and loathsome in the way (as we have seen) an unwilling and guilt-ridden addict or sadist might regard certain of his/her own pleasures.

Incidentally, in speaking of fundamental or basic injustice, I have in mind something like what Rawls means when he refers to the (in)justice of the basic structure of society.

precisely cover up the fundamental injustice of things and thereby serve the interests of the 'ruling classes'. Affection and good manners whose basic social function is to deceive people about their valid interests can hardly be regarded as admirable, and the Marxist might attribute our tendency to devalue them in the face of injustice to an implicit recognition of their ideological function.

But even if we reject these Marxian views, there is a slightly weaker claim that may help to explain why we are inclined to treat civility and community as dependent, for their status as social virtues, on the social virtue of justice. There is something misleading about the presence of civility and community in unjust circumstances, even if (when) they are not actually helping to sustain that injustice. For wherever there is mutual affection or gracious civility, there is, at the very least, the *appearance* that everything is all right, so when these traits exist in a fundamentally unjust society, we have, as a result, the appearance of a good, even ideal, social state of things without the substance, the reality.[7] (If we were to come across a tribe in the jungle and, observing them unobserved, see its members all fraternizing affectionately at some feast or high occasion, we would be surprised—it would seem eerie—if the next day we saw all that familiarity disperse into a rigid caste system of labour and everyday social relations.)

Of course, in saying this I am speaking only about people —not about all possible rational creatures. But something about the moral and emotional repertoire we bring to these matters does make it natural for us to take civility and affectionate relations as a good sign, as a sign, that is, of a good state of things; and so for us at least there seems to be something misleading about an unjust society where civility (gentility) and mutual affection notably flourish. In such

[7] Cf. similar remarks of Kant's in 'Concerning the Social Virtues', from *The Metaphysical Principles of Virtue* (Part II of *The Metaphysics of Morals —not The Groundwork of the Metaphysics of Morals*), Indianapolis: Bobbs-Merrill, 1964, p. 140.

cases, the appearance of good or excellence seems to hide a core-rottenness, and so the factors that create that appearance tend to seem superficial, empty of substance, and for that reason bereft of the validity or merit they might otherwise possess. (A similar explanation, indeed, seems relevant to various earlier-mentioned examples of dependence. There is, for instance, something ironic, eerie, surprising to the likes of us, about mutual trust in the absence of mutual fidelity.) Naturally, this explanation presupposes our own particular categories and tendencies; but claims of virtue-dependence are not supposed to be evaluationally presuppositionless, and our appearance/reality explanation of our belief in (certain) such claims may be a valid explanation even if it is made against a background of moral and psychological assumptions.

III

Are there, then, any social virtues not similarly put into the shade by the glare of injustice? The value of individual virtue in unjust situations has, of course, often been recognized, and we also believe that some of the highest attainments of science, philosophy, art, and literature have taken place against an invidious moral background. But anthropologists and sociologists have long distinguished between culture and society, and it would be odd to speak of the accomplishments (say) of Greek science, philosophy, and art as great social, rather than cultural, achievements. In fact I can think of no clearly *social* virtue not dependent on justice in the way civility and community seem to be; so let me tentatively suggest—though the absence of obvious counterexamples hardly proves the point—that all the virtues of total societies are specifically and unilaterally dependent upon justice—or else are *parts* of justice.[8] (I believe one

[8] The existence of mutual respect among the members of society is certainly a virtue of society, but if a Rawlsian view of the importance of self-respect and, as I understand it, of the connection between self-respect and the respect

can recognize the plausibility of this claim quite independently of accepting any particular definition or theory of social justice.)

If all this is so, then justice occupies a very special place among the social virtues. A being on which all others are supposed to depend for their existence and activity we term a first cause or prime mover, and if all the social virtues unilaterally depend for their status as virtues upon the presence of justice, then it is appropriate to call justice the *first* of social virtues and speak of the *primacy* of justice. Such talk is already familiar from Rawls's *A Theory of Justice,*[9] but when Rawls claims that justice is the first virtue of social institutions, he has in mind something rather different from the unilateral relations of dependency examined above. His idea seems to be that considerations of justice have primacy in the planning and regulation of basic social institutions and are never overriden by considerations, e.g., of (Pareto-type) efficiency or total social well-being. Understood thus, the thesis of the primacy of justice is of a piece with the familiar and widely accepted meta-ethical thesis that the avoidance of wrongdoing takes precedence over all other motives or reasons in the justification of an individual's actions, a view which has recently come in for a good deal of criticism and which I shall be offering my own criticisms of in Chapter 4. But my purpose at this point is not to cast doubt on Rawls's view of the primacy of justice, but rather to set alongside it another conception of primacy with a certain independent plausibility. Rawls's particular

of others is on the right track, then mutual respect is a necessary *element* of social justice—indeed, this seems to be true on a wide variety of conceptions of justice, not merely on Rawls's. (For relevant references in Rawls, see e.g., *A Theory of Justice*, Cambridge, Mass.: Harvard University Press, 1971, sections 67, 76–81, esp. p. 501.) And if it is easier to imagine mutual affection existing apart from justice than to imagine mutual respect thus existing, perhaps that is because the latter is so intimately connected with justice. On the other hand, respect is not admiration, and I doubt whether mutual admiration is any sort of social virtue, as opposed to simply being a good thing when the admiration is deserved or realistic.

[9] See op. cit., pp. 1–10, 31 f., 60 ff., 250, 302 f., 449, 565, 586, e.g.

interpretation of the primacy thesis is intended, among other things, to give expression to our intuitive sense of the special status justice has in ethical conceptions of society, and our own understanding of primacy is meant to reinforce that intuition.[10]

IV

I have claimed that certain virtues and personal goods count for little or nothing in the absence, respectively, of certain other virtues and goods, but it is important to see exactly what this entails. The idea that other social virtues depend on justice does not, for example, entail that they are unimportant in themselves or have little value in comparison with justice. There may be virtues whose importance is exemplified and visible only when certain other virtues are also present, and in a society where a great spirit of community exists alongside just institutional structures, the sense of community and the justice may both reasonably be viewed as great social excellences: indeed I can see no good reason why these two excellences might not be *equally* important aspects of what one prizes and honours in that society. The fact that community depends upon justice and that justice is always and everywhere a virtue may lead us to say that justice is *in overall terms* a more important virtue than community, but that seems compatible with claiming that in certain circumstances where both are present, the spirit of community may shine as brightly (or more brightly) than justice and make an equal (or greater) contribution to what one values and admires about a given society. (If we imagined that sodium was edible and chlorine poisonous, but that they combined to make edible salt, would that entail that

[10] Rawls speaks (op. cit., p. 31 f.) of the priority of the right over the good and takes this to mean (or to be explained by the fact) that putative *individual* virtues (merits) and benefits fail to count as such when (or whenever) they are *incompatible with* justice (what is morally right for society). And this is interestingly similar to the view espoused here according to which certain putative *social* virtues lack that status *in the absence of* justice.

chlorine made a less important contribution to the food value of salt?) Similarly, the belief that mutual trust is rendered empty by the absence of mutual fidelity need not commit us to regarding the latter as more important or valuable *when both occur together*. And surely we may also regard someone's humility as the most remarkable thing about him, even though that trait would have little or no value in our eyes unless the person possessed other desirable traits. So as a general rule the thesis that some particular virtue depends (unilaterally) on another need not commit one to saying that the former is *always* the less important of the two. And the very same point can be made about dependent personal goods.

On the other hand, if one takes too simplistic a view of what good things *are*, one will perhaps be suspicious of the way the notion of dependency cuts across the familiar distinction between the intrinsic and the instrumental. There has, for example, been a tradition of classifying all the goods (ends) of human life (and, among these, all the benefits a person may receive) either as goods worth having or enjoying for their own sake or else as means to such goods; and given the traditional notion of an intrinsic good as something whose goodness is independent of all other factors, the felt sympathy of a friend is neither an intrinsic good for the woman of our earlier example, nor any sort of means to the (intrinsically good) self-respect she wants and needs. Yet it would have value for her in her present situation if only it were accompanied by such respect. Similarly, on the view that sexuality depends on the presence of love for its value, sexual enjoyment is neither an intrinsic good nor necessarily valuable as a means; yet such enjoyment counts as a good thing *in the context of love*, and if the latter idea is not reducible to the usual intrinsic/instrumental distinction, that may only show how impoverished a conception of human flourishing results from restricting oneself to that distinction.

Of course, someone committed to the sole importance of this latter distinction might reject our examples of dependency and insist that whenever I have said that one thing is a personal good (or virtue) only when accompanied by another, we should rather say that the first thing is itself neutral, but can combine with certain goods (or virtues) to create an (even more) valuable organic whole. On such a view, for example, sex is not a good thing, but simply contributes to the value of certain relationships, and community not a virtue, but merely contributory to the social excellence of certain societies in which it exists.

But this insistence comes from the desire to shove all our talk about goods on to the procrustean bed of the intrinsic and the instrumental, and what is quite clear from ordinary thinking about such matters is that we envisage no such necessity in speaking about the various goods and virtues mentioned in our many earlier examples. We normally see no impropriety in saying that community is admirable or praiseworthy in circumstances where justice is present or that sex is a good thing when love accompanies it, see no need in the interests of accuracy to admit that only sex-cum-affection is really good, only community-cum-justice really admirable in a given society. And that, I think, is because when someone says, e.g., that sex is a good thing, he may not be claiming that it is any sort of intrinsic (or instrumental) good, but simply be giving (partial) expression to a view of sex as a good dependent upon love. In our language, at least, dependent personal goods are just as properly called goods as goods that are intrinsic or instrumental, and as we have already seen, this means that our attributions of goodness cannot be completely accounted for in terms of traditional distinctions. Because we need to introduce the notion of a dependent good in order to say everything we want to about the things that are, or are thought to be, good, it turns out—as so often elsewhere in philosophy—that the phenomena we investigate are much more complex than they initially seem.

4
Admirable Immorality

In previous chapters we have investigated some of the ways in which the complexity of virtue and human good eludes the customary philosophical distinctions between intrinsic and instrumental goods and between absolute virtues and vices, goods and evils. But the new distinctions we have worked with represent not a philosophical imposition—not even a justified and necessary philosophical imposition—upon everyday thinking about virtue and human good, but an attempt, rather, to express tendencies of such everyday thinking that philosophers have largely ignored. The ideas of relative good and virtue and of dependent good and virtue advance our ethical understanding of things by clarifying some of our own deepest attitudes and reactions.

But philosophers have not only ignored certain complexities and qualifications of human good and virtue, but have in a sense set virtue and personal good *against* one another through the frequent advocacy of *a priori* theses that put moral limits upon what is to count as a virtue or admirable trait and of other theses that allow considerations of both morality *and* admirability to constrain, in turn, what counts as a personal (life) good. These restrictions have often seemed suspect to non-philosophers, and it will be my main purpose in this chapter and those to follow to show, by a variety of examples and arguments, that common sense is right to reject the proposed limits on what is a virtue or personal good. In the present chapter, I shall argue, in particular, that we should reject the meta-ethical thesis that moral considerations are always over-riding, and recognize the existence of immoral but admirable traits of character, of virtues (excellences) that run counter to morality. Then, in Chapters 5 and 6, I shall consider, respectively, the widespread philosophical view that personal goods always give rise to prima-facie reasons for action, and the Stoic attempt to circumscribe personal (human) good in the light of the virtue of self-sufficiency (autarkeia). And it will be argued, again, that the restrictions on (ordinary conceptions of) personal good that these theses generate cannot in the end be justified. But first to the issue of admirable immorality.

I

It is widely held that (sincere) moral judgements automatically override all other considerations that may occur to moral agents. But in recent years there have been some notable dissenters: Philippa Foot has argued forcefully that even those who care about right and wrong will sometimes put other considerations ahead of morality without subsequent regret or remorse; and Bernard Williams has set out in fascinating detail some cases where a morally concerned individual might consider a given project to be of greater importance, for him, than all the harm to other people that that particular project entailed.[1]

I would like here to strengthen the argument against the overridingness of morality by being more specific than Foot about some of the cases in which people seem deliberately to do what they believe to be wrong without subsequent self-reproach. But I shall also use one of Williams's examples to argue for a possibility that Williams himself never directly considers; for I want to show, in particular, that if (appropriate) moral judgements are *not* automatically overriding, then morality need not totally constrain the personal traits we think of as virtues, and there may actually be such a thing as admirable immorality. But let me first try to say more precisely what the thesis of 'admirable immorality' amounts to.

[1] See Foot's 'Are Moral Considerations Overriding?' in *Virtues and Vices*, Oxford: Blackwell, 1978; and Williams's 'Moral Luck', *Proceedings of the Aristotelian Society*, supplementary volume, 1976, pp. 115–35.

The claim that I hope to justify has to be distinguished both from a very strong thesis that, it would appear, cannot plausibly be defended and from a weak thesis that most of us will find unsurprising, uncontroversial, and uninteresting. The very strong thesis holds that immoral behaviour *as such* may (sometimes) be admirable and would perhaps be congenial to those, like Nietzsche, who regard morality as a symptom of disease or a brake on human excellence. But most of us can find no reason to reject morality in this way, and since the 'very strong thesis' seems precisely to depend on such a dismissive attitude, we are not likely to be persuaded that there is much merit in it.

Consider, by contrast, the view that we may sometimes admire certain aspects of immoral actions or find people admirable for traits whose possession makes them more likely to act wrongly. According to this weaker thesis, we may admire a robber for his daring while deploring his criminal tendencies. But this is not something we would want to call 'admirable immorality'. For what we admire in the robber's act, the daring, can be conceptually prised from its immorality, and even if we believe that daring makes people more likely to act wrongly (say, by exposing them to temptations others would find too risky), we can easily understand what it would be like for a daring individual to have no tendency to exhibit that quality except in good causes.

If, however, certain other cases do *not* permit such a neat separation between what we admire and what is immoral, perhaps we can use them to illustrate an intermediate thesis interesting and controversial enough to be worth defending. But in order to make absolutely clear what I have in mind, I shall need another example, borrowed from Bernard Williams, to set against that of the daring robber. In his symposium paper 'Moral Luck', Williams makes use of a familiar but somewhat fictionalized Gauguin to discuss the important role luck plays in some of the things people care most about in life. And I would like to draw from and somewhat extend Williams's description of Gauguin—while similarly disclaiming historical accuracy—in order to illustrate the (intermediate) sense in which I believe admirable immorality to be possible.

We are all to a greater or lesser extent familiar with the fact that Gauguin deserted his family and went off to the South Seas to paint. And though many of us admire Gauguin, not only for what he produced and for his talent, but also for his absolute dedication to (his) art, most of us are also repelled by what he did to his family. Williams is somewhat hesitant to claim that Gauguin's desertion is morally indefensible,[2] but I believe that we can persuade ourselves of the wrongness of that desertion and that we can do so without losing our sense of admiration for Gauguin's artistic single-mindedness. Single-minded devotion to aesthetic goals or ideals seems to us a virtue in an artist; yet this trait, as we shall see, cannot be understood apart from the tendency to do such things as Gauguin did to his family, and so is not—like daring or indeed like Gauguin's own artistic talent—merely 'externally' related to immorality. Here, then, we may have admirable immorality in the intended sense of our discussion.

II

In order to see whether this is so, let us first examine the reasons for saying that Gauguin acts wrongly in deserting his family. Then we must consider whether the assumption that he acted wrongly forces us to abandon or disown our admiration for his all-consuming devotion to his 'project'.

Williams points out that Gauguin may not be able to justify his behaviour to the family he abandons, unless they are somehow willing to sacrifice themselves on the altar of his art. They will be naturally inclined to feel themselves

[2] In the original version of 'Moral Luck', Williams remains neutral on the question of Gauguin's moral justifiability; but in a revised version that appears as the title essay of a collection of his recent papers (Cambridge University Press, 1981, p. 39), Williams says that we should 'perhaps' accept the conclusion that Gauguin lacks a moral justification.

abused and mistreated, however his project turns out, whatever its benefits for the rest of mankind. More importantly, perhaps, it will also be difficult for Gauguin to justify his behaviour to himself in any recognizable moral terms. Williams imagines Gauguin to be full of remorse for what he feels he 'must' do, to have a powerful and bitter sense of the harm he is causing to those he loves. And indeed any attempt to relax this assumption would tend to undermine our admiration for his artistic passion or impulse—as opposed to his talent (genius) or his *œuvre*. If, for example, we imagine Gauguin offering himself some sort of (Act-)Utilitarian moral justification for leaving his family (or for not feeling remorse about leaving them), we shall, for one thing, suspect that he is simply casting about for excuses to leave and doesn't really care about them. But then there will be no reason (not just epistemically, but in the facts of the case) to suppose he needs anything like passion in order to leave them, and, because of his egocentricity, it will be natural to characterize him as devoted to his own career or his own talent rather than to some more impersonal and objectively valuable conception of what needs doing in art.

On the other hand, we could try to suppose that he was genuinely convinced by Utilitarian arguments into a lack of remorse about what he was doing (had done), but once again we thereby seem to give the passionateness we find so admirable in him no foothold in the facts. For if he loves his family, he will be specially concerned about their well-being; and if for Utilitarian reasons he lacks remorse for what he is doing, then the claim that he loves them seems irretrievably undermined. (Indeed, it is hard to see how a complete commitment to Act-Utilitarianism is compatible with love for particular individuals, with all the special concerns that implies.)[3] In that case, there will be nothing substantial for his passion to overcome, and so nothing in the previously described facts of the case to make it a good example of

[3] The point is made by Michael Stocker in 'The Schizophrenia of Modern Ethical Theories', *Journal of Philosophy* LXXIII, 1976, pp. 453–66.

admirable artistic single-mindedness. It will be easy for Gauguin to leave his family.[4]

However, Gauguin's passionate devotion to his project did not simply make it more likely that he would mistreat his family; rather, it is criterial of having a passion that incompatible impulses, concerns, desires, tend to give way to it, that one is, in effect, *driven* by the passion.[5] Someone passionately, single-mindedly, devoted to art can be expected, among other things, to give less than usual weight to his own safety, to prudence (Gauguin, travelling in days of comparative unsafety, is perhaps a good example of such an attitude). And by the same token, the case for artistic passion may be thoroughly undercut if an artist isn't willing to slight the welfare of others, even of his own family, in its interest. So it seems a condition of Gauguin's recognizable possession of (admirable) artistic single-mindedness that we not imagine him (more or less successfully) arguing himself out of the remorse it would be natural for any moral being to feel in such circumstances; and yet if Gauguin ought to feel remorse, it is for making a choice endemic to the very passionateness with which he embraces his artistic goals.

Of course, Utilitarians and others will want to distinguish what Gauguin ought to be feeling from the question whether he does the right thing.[6] But even if Gauguin's leaving his family was optimific, only the most extremely objectivistic Act-Utilitarianism will say that that settles all important moral issues about the case. Some Act-Utilitarians, for example, claim that one acts rightly only if one does what maximizes the *probable* well-being of humanity or all sentient creatures. Since Gauguin was running a great risk of

[4] Of course, passion *may* exist unopposed; I am merely saying that good illustrative examples of passion(ateness) do not.

[5] Perhaps that is (part of) why it is so difficult to imagine anyone being (passionately) in love with more than one person. (Compare the difficulties in imagining two perfect beings.)

[6] For such others, see Michael Walzer's 'Political Action: The Problem of Dirty Hands', *Philosophy and Public Affairs* 2, 1973, pp. 160–80 and Bernard Williams's 'Politics and Moral Character', in S. Hampshire, ed., *Public and Private Morality*, Cambridge: Cambridge University Press, 1978, pp. 55–73.

never even getting to Tahiti (and of succumbing to sickness or death once he got there) and since the production and reception of works of genius are notoriously unpredictable,[7] it seems that Gauguin could not reasonably have counted on maximizing utility by leaving his family. So an epistemically-qualified Act-Utilitarianism will not be able to say that Gauguin acted rightly in doing so. But even those objectivists who tie the rightness of actions to actual results may none the less (like Smart) insist on a distinction between the rightness of actions and actions it is right to praise, or (like Moore) hold that one may be morally to blame for doing what is in fact right.[8] So I think most Act-Utilitarians (and all Rule-Utilitarians) will have little reason to claim that Gauguin is morally blameless, despite the results of what he did.

But, of course, what most of us find morally objectionable, even repelling, in Gauguin's behaviour is best put in terms altogether outside the Utilitarian scheme of things. Many of us believe that people with families have special obligations to them, that charity begins at home. And even those who believe in an even-handed treatment of family and strangers seem loath to allow that domestic cruelty or unconcern—the total abandonment of a family—can be justified by splendid public performances.[9] Moreover, our tendency—and I believe it is a very deep tendency to be found even among Utilitarians[10]—to accord the prevention of suffering a greater moral weight than the creation of satisfactions, would also make us doubt the morality of what (we are imagining) Gauguin did to his family. All we have said, then, suggests the wrongness of Gauguin's desertion, and so the principal

[7] A point made by Williams in 'Moral Luck'.

[8] Moore, *Ethics*, London: Oxford University Press, 1963, pp. 113–21; and Smart in 'An Outline of a System of Utilitarian Ethics', in Smart and Williams, *Utilitarianism: For and Against*, Cambridge: Cambridge University Press, 1973, *passim*.

[9] e.g., more or less explicitly Utilitarian theorists like Peter Singer in 'Famine, Affluence, and Morality', *Philosophy and Public Affairs* I, 1972, pp. 229–43.

[10] Cf. Singer, op. cit. and Karl Popper's 'Negative Utilitarianism', in *The Open Society and its Enemies*, London: 1974, vol. I, ch. 9, n. 2.

issue before us is whether we can properly accommodate our initial tendency to admire Gauguin's artistic single-mindedness to this conclusion.[11]

III

Our problem arises, with fullest force, out of the common philosophical conviction that morality automatically overrides all opposing considerations—more precisely, that there cannot be any (overall) justification for doing what is morally wrong. If we hold such a view of morality, we shall want to say that someone who does serious wrong should afterwards have a sense of alienation from (and commensurate remorse/regret about) what he has done, and in fact *will* feel this way if he is not beyond the reach of morality altogether. Such a person, we shall agree, cannot feel *justified* in acting as he has and will be unwilling to *stand by* what he has done. But in that case, we shall have a hard time understanding how someone could properly admire Gauguin's passion for art, while holding what he did to be morally wrong. For we have argued that that passion is intrinsically connected with the

[11] Thomas Nagel's reply to Williams in the symposium 'Moral Luck' suggests one further moral defence of Gauguin. As I mentioned in Chapter 2, Nagel holds that luck can play a role in our moral evaluation of actions—even to the point of changing an evaluation from positive to negative. Thus—to revert to our earlier example—if a sentry on (reasonable) orders to kill strangers on sight does not kill an approaching pregnant woman with a toddler in tow, he may be commendable, if the woman is as she appears, but condemnable, if she turns out to be a spy. And so it might seem possible to argue that Gauguin's work in Tahiti, and its subsequent reception in the art world, morally vindicate his decision to abandon his family, a decision that would, however, have been totally wrong, if his career had been less 'lucky'. But there are important disanalogies here from the case of the sentry. For the latter does only one thing—decide to let the woman and her child live—so if our attitude towards (what) the sentry (does) depends partly on what happens later, our variable evaluation has only one place to focus: his one act will be seen favourably or unfavourably depending on what actually happens. But Gauguin not only leaves his family but does subsequent work as an artist, and rather than say that his 'good luck' determines the moral 'sign' of his original decision to leave, we can more plausibly (though perhaps not very plausibly) say that his subsequent work does something to make amends for his earlier wrong.

tendency to do such things as Gauguin did; and since it is difficult to see how one could properly admire the doing of things that cannot be justified and that justifiably arouse a sense of remorse and self-alienation in the agent, it will be accordingly difficult to see how a tendency that intrinsically involved the doing of such things could be a virtue or admirable.[12] From that standpoint we ought to deplore and disavow whatever admiration we do tend to feel about Gauguin's passion for his art—and at most admire his talent/genius or his paintings (as detached from the means of their production).[13] But what if morality is not overriding in the above sense?

I spoke earlier of Philippa Foot's contention that people who 'care about' morality may none the less sometimes allow other considerations to override morality and do what they agree to be wrong without reproaching themselves subsequently for their own moral weakness. Now one of Foot's examples concerns people who spend money on frivolities, acknowledge that it is wrong to do so in a world of hunger and poverty, and yet never 'lose a wink of sleep'. This phenomenon is, of course, familiar; but I wonder whether it should discomfit those who believe immorality overrides all other considerations. The latter may doubt whether the people in Foot's example really hold the moral convictions they claim to hold and they can point to the self-acknowledged frivolity of what those people spend money on in support of this. Furthermore, the fact that the people in Foot's example admit their own frivolity indicates that they do not feel justified in acting as they do. This itself may indicate a sense of remorse about, or alienation from, what they have done, even if they lose no sleep over it, but, perhaps more importantly, it also demonstrates their unwillingness (inability) to

[12] A complete dichotomy between what is admirable and what is wrong is (for example) suggested in several places in Richard Brandt's *A Theory of the Good and the Right* (Oxford University Press, 1979). See especially pp. 166 ff. and 288 f.

[13] Williams never suggests Gauguin's passionateness as a focus of our admiration; he speaks, rather, of 'saluting' Gauguin and his work.

stand by their wrongdoing. And we need an example where an agent feels justified in doing wrong, and is willing to stand by what he does, in order to undermine the 'overridingness' thesis.

However, Foot alludes to other cases in which (unlike the frivolous spender) a person feels he *must* do something he believes to be wrong. The 'must' here is naturally thought to express a sense of justification—rather than (merely) of physical necessity or psychological compulsion—so if there really are cases where someone feels he must do what is wrong, cases that do not give rise to the suspicion that the person in question lacks the moral conviction he says he has, the thesis of overridingness may be undermined. Unfortunately, Foot says no more about such cases than that they involve someone acting in order to 'stave off disaster to himself, or his family, or his country'. But I believe that we can be more specific and that further detail at this point may usefully strengthen Foot's case against the overridingness thesis.

A father may deliberately mislead police about his son's whereabouts, even knowing that the son has committed a serious crime and even while acknowledging the validity of the local system of criminal justice. He may feel he mustn't let the police find his son, but must, instead, do everything in his power to help him get to a place of safety, even though he is also willing to admit that there can be no *moral* justification for what he is doing. Parental love can lead someone to act this way and allow him to feel justified in doing so, and, as I indicated, that feeling may express itself in the thought that he simply has (had) to help his son escape. But even if it can sometimes be right to give preference to members of one's family or to lie to someone bent on harm, valid morality cannot reasonably be thought to include a principle allowing one to lie to legitimate authorities in order to save a guilty offspring. And in any event, there is no reason to deny that the devoted parent of our example sincerely believes that the moral injunction against lying allows no exception in the case at hand. The cause in which he acts is sufficiently important from his own standpoint, that one can see how it might outweigh something else that *really counted with him*. So the fact that he opts against morality—and is willing to stand by what he has done—cannot plausibly be taken to show that his previous concern for (the) morality (of truth-telling) wasn't genuine. (We may also imagine the father to feel some consternation and anxiety about deceiving legitimate authorities.) We thus seem to have found a case where certain (personal) considerations *override* the (believed) fact of immorality, a case where the factors that mar Foot's frivolous spending example are precisely absent.

I am not, however, claiming that every loving parent will decide to act this way. Even if certain sorts of passion (by definition) involve a tendency to override the fact of immorality, some kinds of love may not. It is possible to imagine another parent who felt he *couldn't* mislead the police, and such a person would doubtless be understandable to us.[14] But some parents will feel that they must not let an offspring be captured. And they too are understandable to us.[15] But if such people are humanly understandable, it seems distinctly possible that in other cases a tendency to let

[14] Would such a parent necessarily love his son less than the parent who is willing to deceive? Perhaps, but then what does one say about Richard Lovelace's 'I could not love thee (Dear) so much / Lov'd I not honour more'?

[15] In 'Virtue and Reason', (*The Monist* 62, 1979, pp. 331–50), John McDowell has powerfully argued that a virtuous individual does what is right through having a distinctive way of seeing things, a special awareness of how things are. Now it is plausible to hold that an amoralist fails to see certain things the virtuous person does, and when someone does wrong against his better judgement, it is natural to suppose that this must occur through a clouding of his judgement or senses and that the virtuous person who does right sees something the akrates misses. However, in the case where someone who cares about morality prefers the well-being of his own son and is willing to stand by that decision after the fact, the metaphor of clouding seems inappropriate. Perhaps the virtuous person who refuses to mislead the police has a distinctive way of seeing things, but so too, we might say, does the person who puts the welfare of his son ahead of moral considerations. And if it is claimed that the virtuous father keeps the general facts of his situation more clearly in view, it may perhaps be replied that the father who lies keeps a better focus on just what his son means *to him*, on the closeness of *their relationship*.

certain considerations count for more than morality may actually be admirable. Both the devoted parent and Gauguin may recognize that they are doing wrong and yet also stand by what they feel they must do; but such parental devotion is more in the ordinary run of things, less rare, than Gauguin's single-minded devotion to his art; also, its object is purely personal (familial), whereas what Gauguin tried to do is thought relevant, perhaps important, to all of us.[16] These differences may mean that, while both are counterexamples to the overridingness thesis, there is a better case for finding Gauguin's devotion admirable, not merely humanly understandable. But the very clarity of the devoted parent case may, on the other hand, make it *easier* for us to see Gauguin as a counterexample to the overridingness thesis. In addition, the overridingness thesis appears to entail that Gauguin's remorse at leaving his family lacked sincerity, because at the deepest level he never felt unjustified (overall) in doing what he did. And, as we saw earlier, the suspicion of hypocrisy would tend to undermine our ability to admire Gauguin. But the denial of the overridingness thesis undercuts this argument against his sincerity. So the parental example may not only help us to see the case of Gauguin as a counterexample to the overridingness thesis, but also clear away an objection to the admirability of Gauguin's artistic passion.

However, our defence of the idea of admirable immorality also requires us at this point to distinguish Gauguin's passionate single-mindedness from certain other cases that have recently been discussed by philosophers concerned with different forms of Utilitarianism. Thus writers like R. M. Hare and R. M. Adams have discussed the possibility that certain traits of character or motives or convictions might have a tendency to lead people to perform (what Act-Utilitarians would regard as) wrong actions, but that those traits, etc., might none the less count as morally good because they led to greater utility or to morally better behaviour than would be achieved in their absence (or through the relevant alternative

[16] Cf. Williams, 'Moral Luck', p. 133 f.

traits of character, etc.).[17] According to Hare, for example, the firm conviction that slavery is always wrong may very well lead to better results and to morally better behaviour than any less uncompromising (and more Act-Utilitarian) attitude. Those with a pure Act-Utilitarian attitude might well, he thinks, be tempted, through ignorance or self-interest, to condone slavery in cases in which, though slavery was actually harmful, it could be colourfully represented as being optimific. So an uncompromising aversion to slavery might be a good trait (conviction) to possess, despite the fact that, for an Act-Utilitarian like Hare at least, such a trait is inconceivable apart from the tendency to oppose slavery in those (rare or purely hypothetical) situations where such opposition is wrong. Similarly, the Act-Utilitarian may want to hold that parental devotion is a morally good motive/attitude by virtue of being optimific or by virtue of leading to morally better behaviour, of being 'moralific' (if I may be allowed to introduce an analogous term). Yet such devotion is logically connected with a tendency to favour one's children in some circumstances where greater good can be achieved by a more impartial attitude and where, again by the Act-Utilitarian's lights, such favouring would be wrong.

However, the Utilitarian (or other theorist) who took this sort of view of parental devotion or unalterable opposition to slavery would not need to view these traits as running counter to morality, or thus as instances of admirable immorality. For in the examples we have briefly mentioned, a character trait or motive (essentially tied to a tendency to wrongdoing) is said to be good (or admirable) only on the basis of its leading to morally good action or to the overall good results that are the stock in trade of utilitarian moral justification. So someone who regards these traits as good or right can say that if and when a person acts wrongly as a

[17] See Hare's 'What Is Wrong with Slavery?' *Philosophy and Public Affairs* 8, 1979, 103-21 and his *Moral Thinking*, Oxford University Press, 1982; also Adams's 'Motive Utilitarianism', *Journal of Philosophy* LXXIII, 1976, pp. 467-81. I have taken various suggestions about how one may morally justify traits or motives from these sources.

result of possessing one of them, there is none the less a moral justification for the tendency that leads to that action, and the particular action itself can then be said to be justified as an indispensable part of a morally desirable overall pattern.

In that case, the examples of parental devotion and absolute opposition to slavery (as viewed from an Act-Utilitarian standpoint) have force against the overridingness thesis as originally enunciated, by pointing out the possibility of some sort of overall justification for particular wrongdoing. But what in those examples seems to justify given wrongdoings is always 'larger' moral considerations, considerations, in particular, of the overall moral significance of certain traits or motives; and it requires only a very modest modification of our original overridingness thesis to accommodate the idea that the morality of motives or character traits may sometimes override the morality of single actions. We can now say that anyone with a serious sense of morality will feel regret/remorse about doing what he acknowledges as morally wrong, unless he regards it, and can justify it, as a necessary part of a larger and morally justified pattern, as indispensable, e.g., to the overall improvement of moral behaviour or to the general well-being. And in consonance with this modified overridingness thesis, one might also claim that character traits or motives that are logically linked to a tendency to do wrong in particular cases, can be good or admirable only if their overall tendency can be *morally justified*. But then the distinctions recently drawn between the morality of motives and the morality of actions in no way suggest the possibility of admirable traits that *run counter to morality*, and that is why the sorts of cases I mentioned earlier must be brought in if one seeks to defend the possibility of admirable immorality and challenge the thesis of moral overridingness in a more than superficial way.

The examples mentioned in the recent literature of utilitarianism may involve traits with an essential connection to wrongdoing, but the case of Gauguin involves far more than this. It is a case where the essential tendency to particular wrongdoing does *not* seem morally justifiable as part of a larger pattern. Thus parental devotion or absolute opposition to slavery may perhaps be held to be moralific; but nothing similar can plausibly be said about singlemindedness in the cause of art. For if we imagine everyone (or every artist) passionately devoted to an aesthetic ideal or cause, we must presumably imagine a great deal more of the sort of behaviour commonly attributed to Gauguin, and we have already argued in the most general terms that what Gauguin did was morally wrong. By the same token, even if parental devotion (or opposition to slavery) on everyone's part (or on the part of any random individual) can perhaps be said to increase probable utility, it seems utterly implausible to suppose that the possession of artistic single-mindedness by people generally or by an average person would increase the general welfare. (How frequently are artistically brilliant results likely to compensate, in terms of utility, for neglected family and fellow-citizens? And how many great works of art have actually depended, for their existence, on an artistic single-mindedness comparable to Gauguin's? Surely, we have, in terms of utility, more to fear than to hope for from passionate single-mindedness in the cause of art.)[18] Gauguin's single-mindedness is thus, if anything, a morally *un*justified motive or character trait, and any virtue we find in it, any admiration we feel for it, is predominately *not* of a moral kind. (One would have to be an extremely objectivistic utilitarian to claim that Gauguin's single-mindedness is morally good, and admirable, because of the results it led to in his particular case, and such a view, in any event, implies something I think almost everyone would find unacceptable, namely, that Gauguin's single-mindedness is admirable only because he wasn't accidentally killed on the way to Tahiti.)

[18] Given what has just been said, is there even any convincing reason to believe it would be optimific for *all talented artists* to be single-minded in Gauguin's fashion? (Notice the attendant epistemological difficulty of identifying talented artists.)

Thus, much as we find with such traits as high intelligence and physical daring, our admiration for artistic single-mindedness is not necessarily paid in moral coinage; but the latter trait is distinguished from the first two by being inconceivable apart from a tendency to wrongdoing. So, given, in addition, that, in the actual world at least, such single-mindedness also lacks any overall moral justification, given that, by contrast with parental devotion and opposition to slavery, such single-mindedness tends to produce wrong action and disutility, the case of Gauguin seems to provide us with an admirable trait that runs deeply counter to morality, with a genuine example of admirable immorality—assuming always that our sense of what is admirable here remains defensible.

In consequence, even the modified overridingness thesis falls prey to the Gauguin example. For if Gauguin can stand by what he does yet acknowledge its wrongness, it is not because he is in any position to claim an overall moral justification for his tendency to put his artistic ideals before all other considerations. And something similar can be said about the earlier-mentioned case of the parent who lies to legitimate authorities in order to save a guilty offspring from just punishment. I have already suggested that, unlike certain kinds of passion and single-mindedness, parental love may be able to exist apart from the tendency to do the sort of thing the parent in our example does. And in that case his action cannot presumably be morally justified as the inevitable consequence (in his particular circumstances) of a moralific or optimific motive or attitude. But in any case the possibility of such a moral justification may well be the furthest thing from the mind of the parent who deceives or has deceived the authorities. Realistically speaking, what makes him feel justified and willing to stand by what he has done is not (the possibility of) an overall moral justification in terms of the morality of traits and motives, but rather the particular danger to his son. So even the modified overridingness thesis that is suggested by recent discussions of the distinction between the morality of acts and the morality of motives gives an insufficient sense of the limits of the role moral considerations play in our assessment and justification of actions, traits, and motives, and is thus inadequate to describe the actual phenomena of the moral life.

Once, moreover, we cease to see phenomena through the prism of an overridingness thesis and grant that admirable immorality is not automatically spurious or suspect, we may begin to wonder whether there are other plausible cases of the same phenomenon and also whether the idea of admirable immorality is really anything new.

IV

It is worth asking, for example, whether Kierkegaard's 'teleological suspension of the ethical' may not anticipate what I have been saying. Kierkegaard held, in particular, that Abraham's willingness to sacrifice his own son at God's command involved going beyond the ethical. And since Kierkegaard's admiration for Abraham's faith/devotion does not, as with Nietzsche, depend on any general rejection of morality, Kierkegaard seems to be saying that morality must sometimes be overridden in the name of something (even) higher and thus to be treating Abraham as admirably immoral in the sense of our discussion.

But the issue is not so unambiguous. Kierkegaard accepts the Kantian view that the ethical is 'the universal' (that is, operates through universally applicable rules), and he says that what Abraham does, he does by virtue of his particular relation to God. What Abraham does thus cannot, he says, be formulated or enjoined in terms of universal rules, it is beyond the sphere of the ethical. But Kierkegaard then muddies the waters by speaking of Abraham's (having a higher) *duty* to do as God commands. As a result, there seem to be (at least) two possible ways to understand Kierkegaard's position on admirable immorality. We could, on the one hand, interpret him to be saying that morality must be conceived in Kantian terms, so that the fact (if it is a fact) that Abraham goes beyond the sphere of general rules, beyond the ethical, means that what God asks him to do is wrong, but that he is none the less admirable for a devotion that makes him willing to obey God in all things. This reading might lead us to regard Kierkegaard as committed to the idea of admirable immorality in the sense under discussion, but it also forces us to disregard (or treat as Pickwickian) all his talk about a duty to do as God commands.

On the other hand, we might take the talk about duties to God at face value and refuse to assume that Kierkegaard believed Abraham's intended action to be wrong; the point about the teleological suspension of the ethical would then have to be that the moral permissibility and admirability of that action cannot be accounted for in terms of universal (or universalizable) rules, in terms of the ethical as Kant imagined it. I find it impossible to say which of these interpretations is better, and there may indeed be no answer to the question asking which is closer to Kierkegaard's intended meaning. So we are in no position to claim Kierkegaard as an earlier exponent of the admirable immorality thesis.[19]

Of course, we might still consider the case of Abraham in its own right and ask whether it in fact involves admirable immorality. But it may be impossible to consider Abraham's case with the sympathy and appreciation that actuated *Fear and Trembling* without presupposing a more extensive religious framework than most readers are likely to accept. So at this point perhaps it would be best simply to drop the issue of Abraham's admirable immorality. That need not, however, prevent us from discovering *other* examples of this phenomenon.

We have already found one plausible instance in the artistic

[19] Stanley Cavell (*The Claim of Reason*, Oxford University Press, 1979, p. 268 f.) claims Kierkegaard (not to mention Nietzsche) as an exponent of the thesis, but does not consider the objection to this interpretation of Kierkegaard mentioned in the text above. Interestingly, Cavell himself seems favourable to some sort of admirable immorality thesis, but mentions no specific examples on which to base it.

single-mindedness of a freely rendered Gauguin. But certain forms of passionateness or single-mindedness outside the artistic sphere can also appear admirable, and yet (of necessity) exemplify the same sort of tendency to slight the moral that we found in Gauguin. Consider, for example, a Winston Churchill portrayed with similar freedom. In conformity with the popular image of Churchill as a resolute wartime leader, we can imagine him single-mindedly devoted to crushing Nazism, to Allied victory. As a result, late in the War he approves the fire-bombing of certain German cities in the hope of breaking civilian morale and bringing Germany quickly to its knees, even though ultimate Allied victory is by that time reasonably assured. Imagine further (what I believe is historically inaccurate) that the fire-bombing achieves its purpose, rather than simply stiffening German resistance.

Now the bombing of civilian targets goes against the conventions of war, and there is widespread agreement among those who have discussed the issue that there is no moral justification, in the circumstances mentioned, for what Churchill did.[20] Yet it is difficult to see how anyone fitting the popular image of Churchill could have done anything else. The point (and it is similar to what we have already said about Gauguin) is that Churchill's single-minded, passionate devotion to Allied victory would by its very nature impel him to do everything in his power to defeat the enemy (at the cost of fewer Allied lives), and so lead him to do the very things we have just said seem to be wrong. Yet we admire Churchill, and part of our admiration is directed towards the very single-mindedness which led him to act wrongly.

But even if Churchill's single-mindedness was a rare and extraordinary personal quality and we see its object, Allied victory, as a fundamentally good cause, a benefit to mankind generally, it hardly follows that it was a morally good trait

[20] This would be granted even by some who approve the bombing of civilian targets earlier in the war, when ultimate victory was very much in doubt. Cf. Michael Walzer's 'World War II: Why Was This War Different?', *Philosophy and Public Affairs* I, 1971, pp. 3–21.

and can be given a moral justification. For example, it seems easy to imagine a different Churchill who was sufficiently resolute and devoted to Allied victory to do everything necessary to ensure it, without being so passionate and single-minded as to put victory ahead of moral considerations in the way that Churchill seems actually to have done. (By contrast, we have assumed that a merely devoted Gauguin who was unwilling to abandon his family would have been unable to realize his artistic goals.) And if such traits as resoluteness and devotion (or determination and dedication) are likely to result in morally better behaviour than such passionate single-mindedness as Churchill displayed, then the latter hardly counts as moralific. Moreover, even granting Churchill's assumptions about the effectiveness of fire-bombing, it is very doubtful that the sort of decision a passionate Churchill would make in that connection would tend to have greater utility for mankind *on the whole* than what a less single-minded devotion would have dictated. So if (mere) resoluteness, etc. represent alternatives to passionate political single-mindedness, the latter trait not only involves an essential tendency to do what is (generally thought of as) wrong, but also lacks any sort of moral justification as a desirable motive overall. If Churchill's passionate single-mindedness is admirable despite its countermoral tendencies, we have yet another example of admirable immorality.

But, of course, the contrast we have just drawn between (mere) resoluteness, devotion, etc., on the one hand, and passionate single-mindedness, on the other, might at this point lead one to suspect that what we really admire or ought to admire about Churchill is his resoluteness, devotion, dedication, etc., characteristics which could have led to victory without (*ex vi terminorum*) inclining him to do wrong things, rather than his (possibly deplorable) single-mindedness. But perhaps this doubt can be met by insisting, or suggesting, that we can properly admire both (Churchill's) morally neutral dedication, resoluteness, etc., and (his) not-so-neutral single-mindedness in an important and good cause.

I think most people's initial inclination is to admire all of these characteristics (of Churchill's), and I believe only someone under the sway of the overridingness thesis is likely to think that it is really only the attendant resoluteness and devotion, not the single-mindedness itself, that we ought to admire in Churchill. Once, however, the overridingness thesis is called into question, we can perhaps go back to trusting our initial inclination to regard Churchill's single-mindedness in the Allied cause as a virtue, and treat it as another genuine instance of admirable immorality.

Finally, I should mention one further political example, which, because it presents a different focus of admiration from any encountered so far, usefully widens our spectrum of plausible instances of admirable immorality.

A great deal has recently been written on the problem of 'dirty hands', but one particular example of the phenomenon has special significance for our discussion. Michael Walzer has described a (hypothetical) highly-principled political leader who has just taken power in a situation of civil strife in which rebels have placed a number of time bombs around his country's capital city. Government forces have captured one of the ringleaders of the rebellion, who, though he is not responsible for placing the bombs, refuses to give away their whereabouts. Only the ringleader can tell where the bombs are in time for the leader to do anything about them, and in order to save many lives, the new leader authorizes torturing the man, thereby (let us assume) succeeding in getting the required information and preventing a disaster. But this in fact goes against the high principles he has hitherto espoused. (We may imagine him to head a reformist government that has gained power against a regime that routinely used torture.) The new leader may thus knowingly be doing something that he himself considers wrong, for he believes that the use of torture for political ends is always wrong; and for that reason he will need a strong stomach to do what he feels he has to do in order to avert a tragedy. He will need a strong stomach because he knows what torture means—his knowledge is part of his reason for thinking it always wrong. So he will have to conquer his own personal and moral aversion to torture, in order to use it.

Walzer himself wants to say that what the leader does is morally wrong. And yet he also claims that if we value the result, the saving of innocent lives, we may want just this sort of political leader: someone who, while possessed of conscience, can and will overcome that conscience to avert human tragedy, a leader who is willing to get his hands dirty for this sort of reason, in this sort of way. (Cf. Machiavelli: 'Thus it is well to seem merciful, faithful, humane, sincere, religious, and also to be so; but you must have the mind so disposed that when it is needful to be otherwise you may be able to change to the opposite qualities. And it must be understood that a prince, and especially a new prince, cannot observe all those things which are considered good in men, being often obliged, in order to maintain the state, to act against faith, against charity, against humanity, and against religion.')[21]

Walzer points out that we not only honour the results of torture in the case he imagines, but praise the leader who has effected this result. But he does not mention that we also admire the leader precisely for having the will and ability to overcome his personal aversion to torture and do what he takes to be wrong in order to save lives. Of course, our admiration may also focus on his (single-minded) dedication to the public good or on his resoluteness in the public interest, and this would provide a common element with what we have already said about Churchill and Gauguin. But what makes the case different from those earlier ones is that we also admire the new leader's ability and will to serve his people in circumstances that make it extremely difficult and unpleasant to do so. This admiration is not directed at his having, simply, a strong stomach: an ability, say, to contemplate horror movies and excrement alike with equanimity. Strong stomach of this general sort is clearly morally

[21] *The Prince*, Ch. 18: In what ways princes must keep faith.

neutral, but it is difficult to see what is particularly admirable about it, and there is in any case no reason to think that in admiring Walzer's leader, we are assuming he has a strong stomach in this sense. He may not. But if he has the ability to overcome personal aversion and pangs of conscience in a cause of the sort mentioned, then he has what we may call 'moral stomach', and we find this trait admirable for its rarity and value. And what is thus admired cannot be separated from, indeed, is defined in terms of, the tendency to act against conscience and aversion.

But perhaps the leader acts from a mistaken sense of right and wrong and, despite the opposing dictates of his conscience, actually does the right thing in ordering the use of torture to save lives. Such a conclusion may not, indeed, follow from the natural assumption that it would be wrong of him *not* to do what was necessary to save those lives, since it is possible to hold that the leader is, through no fault of his own, in a position where he cannot *avoid* acting wrongly.[22] But perhaps (despite the complications of precedent effects) some sort of Utilitarian argument could be given for the rightness of using torture in his circumstances, and in that case it becomes unclear whether we have a genuine case of admirable immorality.

The point, however, is that it takes Utilitarian or Utilitarian-like assumptions to deprive the leader example of that status. If acts of torture can be morally justified by their results (in this sort of case), then the leader acts rightly and his moral stomach may well be both moralific and optimific and thus count as a morally good trait despite the leader's own view of the matter. But given the state of philosophical discussion of these issues, there is some reason to believe that political uses of torture may be morally wrong even when necessary to save many lives, and consequently that moral stomach as we have defined it is a highly *anti-moralific* trait. But then unless, again, one relies on Utilitarian-like assumptions, it is difficult

[22] See, for example, Walzer, 'Political Action . . .' and T. Nagel's 'War and Massacre', *Philosophy and Public Affairs* I, 1972, pp. 123–44.

to regard moral stomach as a *morally* good trait, so there seems to be a possible and not easily dismissible anti-Utilitarian standpoint from which what the leader does is wrong and lacks any sort of (overall) moral justification. Yet one who finds this standpoint plausible may none the less admire the leader's willingness and ability to do wrong in the service of a great humanitarian good, and I believe Walzer is a likely example of such an attitude. And it is no longer possible to invoke the overridingness thesis to show that this attitude must somehow be incoherent or unreasonable. So relative to certain anti-Utilitarian assumptions, we have yet another example of admirable immorality, a case where what we think of as a virtue (in a leader) cannot be conceived apart from a tendency to wrongdoing and is not susceptible to any overall moral justification. And the leader's own attitude then represents another counter-example to the modified overridingness thesis.

However, there is one further challenge to our defence of admirable immorality that we have not yet considered and that at this point deserves our attention.

V

Some political figures openly enrich themselves at the public's expense. Such corruption is so unsurprising and inevitable that humour or quiet resignation may well be a more appropriate response to hearing of specific instances than any sort of moral indignation. But it hardly follows that one should *admire* the person who gets a reputation for successfully defrauding the public. And yet many of us who care about morality and disapprove of political corruption none the less do, occasionally, catch ourselves admiring certain political leaders for their brazenness, for their ability publicly to mulct the public.[23]

There is surely something suspect about such reactions. It

[23] A point made by Williams in 'Politics and Moral Character', loc. cit.

may be wondered, in particular, whether such (momentary) admiration is not simply transformed envy, the expression of our own wish to get away with what certain politicians clearly do get away with. If so, our admiration may represent a kind of internal akrasia—like the impure thoughts some people fear and feel guilty about. And there would be reason to treat our tendency to admire brazenly corrupt politicians as something to be got rid of or at least disavowed. The tendency towards such admiration would be something that (at the very least) we did not admire in ourselves. And it is some evidence of our actually having such an attitude towards our own occasional outbursts of admiration for the brazenly corrupt that we naturally speak of having a 'sneaking admiration' for them. Just as a sneaking suspicion is one we would rather not (have to) have, because it indicates a state of affairs we would rather not have obtain, so sneaking admiration is admiration that (slightly) embarrasses us, that we would rather not have. And if there is reason to disavow our admiration for corrupt political figures, there is no reason to regard their open corruption as a form of admirable immorality. May there not be similar reasons for suspecting our admiration of Gauguin, Churchill, or Walzer's leader, reasons based not on any *a priori* thesis about moral overridingness, but on the particular nature and sources of the admiration elicited by these examples?

What may give us reason to suspect our admiration for Gauguin's passionate devotion to art, for example, is the fact that there exists (or has existed) a powerful Romantic 'myth of the creative artist' that might be causing or influencing that admiration. According to the Romantic myth, artists are, among other things, a special breed not subject to normal human constraints, in touch with a reality denied to others, and, because somehow closer to the divine, endowed with (nearly) divine prerogatives. If we suspected that, in admiring Gauguin's passion, we were succumbing to this extravagant picture of 'the artist', then we might have strong doubts about the admirableness of Gauguin's passionate devotion—and because of the similarity of focus in the case of Churchill, that example too might be put into jeopardy.

But I find it difficult to believe that we are thus subject to the myth of the artist, that this myth needs to be exorcized like some adolescent 'crush', if we are to put Gauguin in proper perspective. Many of us (no doubt with great benefit of hindsight) now think the Romantic myth escapist and sentimentalized, in parts even frivolous and absurd; and if we are not deceiving ourselves, then this rejection of the Romantic myth can *coexist* with our admiration for Gauguin's artistic single-mindedness. Moreover, it earlier seemed to be a condition of our fullest admiration for Gauguin's passion that Gauguin take morality and his family seriously enough to feel concern and remorse about what he did.[24] So the particular worry we have about our admiration for politically corrupt leaders—that it stems from an akratic abrogation of moral concern, perhaps even from a temporary mocking attitude towards the felt constraints of morality—is irrelevant to our admiration for Gauguin.

But let me also suggest a more positive way of (re)assuring ourselves about any tendency we may have to admire Gauguin, Churchill, and/or Walzer's political leader for the wrongness-involving characteristics we have mentioned. It seems reasonable to explain why we accept our own admiration for Churchill *et al.*, but feel uncomfortable with any admiration we feel for the brazenly corrupt in terms of the aims or purposes involved. For all its connection with moral wrong, the passion we admire in Churchill and Gauguin is directed towards larger, impersonally valuable goods. And this explanation of our differing reactions takes sustenance, I think, from the fact, noted earlier, that self-accepting high admiration for Gauguin's passion in part depends on

[24] It doesn't similarly seem a necessary condition of our admiration for Churchill's single-mindedness, perhaps because it is more humanly understandable that a person should lose sight of his obligation to enemies bent on his destruction. (If Gauguin feels a remorse that, I assume, Churchill never felt, that may make the latter also appear more single-minded.)

conceiving him as passionately devoted to the realization of a certain project, rather than to the advancement of his career or reputation. For only thus will his passion, again, seem directed towards something publicly, impersonally, valuable that people can benefit from.

Our explanation is also supported by the fact that if we imagine passion directed to a *less valuable* impersonal end, e.g., the founding of an art collection rather than the creation of art, it becomes more difficult to hold on to our sense of something admirable in the face of actual wrongdoing done in the name of that passion: an art collector who abandons his family to pursue *that* passion seems monstrous or risible; it seems impossible to compare such a person with Gauguin.[25] Thus differences in the value and impersonality of what is aimed at plausibly explain our differing attitudes towards these cases. And perhaps they also provide a justification for those differences of attitude. Given, in other words, that we are inclined to accept our admiration of Gauguin's and Churchill's passion because they are directed towards great goods beyond the confines of their narrowly conceived self-interest and given that the lack of such concern is, in other cases, what makes us embarrassed by our admiration or unable to feel admiration at all, may we not also plausibly claim that this difference in aim and object is precisely what justifies our different reactions? We could then conclude that Gauguin's and Churchill's passion really are admirable.

What may further incline us in this direction is another sort of case that may have seemed conspicuously absent in the light of those we *have* been discussing. For all our emphasis on artistic and political passion, we have not yet said anything about the very paradigm of passion, sexual love. The omission has been deliberate and the reasons behind it may be instructive. If we seek or invent examples

[25] We also don't want the person passionately devoted to (his) art to overestimate his own talent. Otherwise, his single-minded behaviour will seem more an expression of pathetic delusion, or megalomania, than of admirable devotion to an artistic project.

of sexual passion attended with moral costs—perhaps a suitably imagined Tristan and Isolde will serve—I think it is actually somewhat difficult to evoke admiration of the kind we feel in connection with Gauguin's passion or Churchill's. And the explanation, once again, may lie with the hypothesis defended above.

In order to admire the passion of a Tristan and Isolde, I think we must import and insist upon artistic or political metaphors that imply the objective or public value of their love. For example, if we could think of Tristan and Isolde as establishing a beautiful community of two which all real communities can aspire to (or take consolation from), it would be easy to find their passion admirable, and not merely understandable. But I doubt whether such a metaphor will have much appeal to most of us. Alternatively, one might liken their love to a work of art, and if one were sufficiently struck by *that* analogy, one might indeed feel admiration for their passion. But, once again, it seems difficult to take up such an attitude. Questions like 'Would they really have lived happily ever after?' (more soberly, 'Would they have always wanted to stay together?') are too likely to intrude upon our inclination (e.g.) to treat their brief love as possessing a lasting beauty.

Thus admiration for romantic passion seems naturally to reach out for support from political or artistic imagery, and if our (in)ability to feel admiration for Tristan and Isolde's passion is plausibly explained in terms of our (in)ability to see them as achieving some (political or artistic) value beyond the immediately personal, that very fact should help to allay our doubts about cases, like those of Gauguin and Churchill, where passion clearly *is* directed towards publicly valuable objectives. (In effect, then, the artistic and political cases achieve genuine admirability in metaphorical passion, but literal passion achieves only doubtful admirability via metaphors of artistic or political accomplishment.)

But one more doubt must be mentioned before the case

for admirable immorality can be considered fully made, a doubt that arises from our desire to defend this notion within a framework that allows for the validity and importance of morality. Could it not be argued that if we do care about morality, we must at least have some negative feeling towards a Gauguin whose artistic single-mindedness led him to do things we are granting to be wrong? And if this point is allowed, will it not be difficult to see how genuine and acceptable admiration can coexist with such feeling?

I think it must be admitted that Gauguin's actions should arouse negative feeling in us; but I think such feeling will only be held incompatible with legitimate admiration if one takes a simplistic view of the nature and interactions of emotions. We may indeed be chilled and repelled by what Gauguin does to his family, but the appropriateness of these emotions need not drive out or make inappropriate our admiration for his passionateness. Indeed, the existence of *all* these reactions may be a sign that one is properly appreciating what Gauguin was really like.

Consider a parallel. Some beauty is cold (the examples of Dorian Gray and Anita Ekberg spring to mind); and those who fully respond to the nature of such beauty may at the same time be chilled and repelled by it. But a tension between attraction and repulsion may be absolutely appropriate to such cases, and those who see (and feel) only the beauty may be thought to lack an accurate sense of the particular beauty these people have. A similar tension between admiration for Gauguin's passion and repulsion at what it led him to may be precisely what the case of Gauguin, as we have described it, calls for. Nor is it any objection to such an analogy that passion must be hot, but beauty can be cold. For passion has a direction, and in the direction of its 'proper object' it is indeed 'hot'. But heat in one direction can create what looks like coldness in every other direction, and a passion for artistic accomplishment can thus seem a coldness to those outside the scope of that passion. Gauguin's behaviour to his family (for all his subsequent remorse) naturally chills and repels us, but, as with the case of cold beauty, there is no reason to think that the coldness and the repellingness of his passion makes admiration out of place. Perhaps we cannot fully appreciate Gauguin and what he did without all these feelings, or reactions, without a sense of tension and complication about his case.

It is worth contrasting the above examples with cases of ambivalence where *irrational* emotional tension exists. Where love and hate exist towards one and the same object—and we are in a position to stand back somewhat from these feelings —we are willing to admit that at least one of the feelings is inappropriate. We agree that love/hate beclouds our appreciation of its object and that a better understanding of that object would require us to overcome our ambivalence. Similarly, in cases of unwilling addiction, the desire, say, to have a certain drug exists in tension with the desire not to give in to that desire, and our judgement is clouded, our attitude (as we say) inconsistent: no one expects a person to have a proper understanding of the attractions and disadvantages of drugs in such moments.

But not all cases of emotional tension need be considered irrational and inconsistent, need deprive us of clear-sightedness. Some eighteenth-century views of the sublime required an emotional attitude of mixed pleasure and pain, a simultaneous sense of inferiority and upliftedness, for its appreciation; and we are accustomed to hearing from theologians that a proper appreciation of God (of the 'numinous', or the *mysterium tremendum*) requires an attitude of awe compounded of fear, love, and, dare I say it, admiration.[26] We may no longer find these particular theories attractive, but they at least show that there is some tradition in the idea of emotional tension as required by, rather than preventive of, proper appreciation of some person, quality, or situation. In cases like that of Gauguin, or Churchill, or even Walzer's political leader, there is certainly emotional tension in the attitudes I have agreed are appropriate to certain of

[26] Cf. Iris Murdoch's *The Sovereignty of Good*, London: Routledge, 1970, p. 81 f.

their character traits. But there is no need to characterize that tension as amounting to ambivalence, if by that one means an irrational and inconsistent attitude of clouded judgement. The closeness of the analogy with cold beauty makes it seem natural and appropriate to feel both admiration and a sense of repelling coldness about Gauguin's, or Churchill's, single-mindedness, and someone who has such a mixed reaction to that single-mindedness may actually see it better than those who reject either element of the compound. Thus there is no reason to suppose that a mixture of attitudes automatically indicates that something has gone wrong, and we have found no reason to deny our sense that in cases like that of Gauguin, Churchill, and Walzer's leader, we have found what may properly be called admirable immorality.[27] The common philosophical assumption that morality sets absolute limits to human virtue or excellence seems thoroughly unjustified.

The whole plausibility of the idea of admirable immorality depends, of course, on a loosening of our attachment to the 'overridingness' thesis. But, in addition, there has in recent years been a growing consensus that Hume was right and Kant wrong about the morality of acting from motives other than respect for the moral law (or a sense of duty). The person who visits a friend in hospital from sheer sympathy may not be *morally* superior to one who does the same thing from duty, yet most of us would grant that he does not act wrongly and would be inclined to think better of him than of someone who treated friendship as if it were a matter of acting on duties (of friendship).[28] But if the fact of immorality is not always an overriding consideration for us and if we can think better of someone for acting without attention to right and wrong, is it so very surprising that we should sometimes see virtue in traits that actually run counter to morality?

[27] Gauguin will have, if not logical reasons, then at least reasons of modesty for not admiring his own passionate single-mindedness, but he certainly may be chilled by what he has done—what *he* has done—to his family.

[28] The example is from Stocker, op. cit.; see also the title essay of Foot, op. cit.

5
Goods and Reasons

As we have just seen, philosophers have often put unjustified moral limitations on what may count as an admirable trait or virtue; but sometimes considerations of morality, admirability, or virtue are themselves in turn used to restrict the class of personal life goods. For example, it has been held that in circumstances where such (otherwise) good things as power, position, wealth, and pleasure can only be obtained by flouting moral requirements, a virtuous individual loses nothing by not seeking them. The pleasure or wealth that is thereby forgone is not a good outweighed by the goods attainable through or in acting virtuously, but is no sort of good whatever.

For all its initial counterintuitiveness, this view has recently been powerfully defended: both in its own right and as an interpretation of Platonic and Aristotelian ideas about human flourishing. In what follows I hope to show, however, that the view rests on a vulnerable assumption about the relation between goods and reasons. And once we deny the need to relativize the good of wealth, power, etc., to individual circumstances, we shall have to reconsider the common philosophical opinion that the inherently 'shameful' pleasures, e.g., of sadism and drug addiction are not even momentary personal goods for those who enjoy them. It will turn out to be a distinct possibility that power, wealth, pleasure, and the like are *always* good.

I

The case for relativizing goods to individual circumstances proceeds from a widely neglected moral-psychological insight. In a series of papers, John McDowell has argued that the virtuous individual's perception of moral requirements silences all opposing considerations.[1] Such a person does not weigh the pleasure or money obtainable by an unjust act against the pleasures of acting virtuously and/or the shamefulness of injustice and then refrain from acting unjustly on the basis of the balance of reasons. Rather, on those occasions when pleasure or money cannot be honourably attained, they offer the virtuous (as opposed to the merely continent) individual *no reason whatever* for doing what is contrary to virtue, to 'requirements of excellence'.

A similar point is made by Gary Watson in his article 'Free Agency'.[2] Watson discusses a possible individual who believes his sexual inclinations are the work of the devil and accordingly places no value on obtaining sexual pleasure. And he holds that such a man will regard his sexual inclinations as giving him no reason for indulging in sexual activity. A man of this sort stands in marked contrast, he thinks, with someone who believes something can be said in favour of sex, believes sex has some value, but holds that on balance total celibacy is better. The latter has, or regards himself as having, a reason for sexual activity that is outweighed by other considerations; the former will not acknowledge even a prima-facie reason for engaging in sexual activity.

Having each made the important distinction between the outweighing of prima-facie reasons and the total silencing, or nullification, of reasons for a given action,[3] McDowell and Watson take a further step which they regard as innocuous. Relying on what he takes to be the 'formal' intertranslatability

[1] See 'Are Moral Requirements Hypothetical Imperatives?', *Proceedings of the Aristotelian Society*, Supplementary Volume 52, 1978, pp. 13-29; 'Virtue and Reason', *The Monist* 62, 1979, pp. 331-50; and 'The Role of *Eudaimonia* in Aristotle's Ethics', in A. Rorty, ed., *Essays on Aristotle's Ethics*, Berkeley: University of California, 1980, pp. 359-76.

[2] *Journal of Philosophy* 72, 1975, pp. 205-20.

[3] W. D. Ross (in *Foundations of Ethics*, Oxford: Clarendon, 1939, pp. 108 ff.) makes a similar distinction between what overrides (outweighs) and what cancels (abolishes) a prima-facie duty (releasing someone from a promise is an example of the latter). But such a distinction has application only within the sphere of moral reasons, whereas McDowell and Watson are thinking of cases where moral requirements nullify, or cancel, non-moral reasons for action.

of talk about reasons and talk about goods, advantages, and losses, McDowell infers that the virtuous man or woman who forgoes/ignores a pleasure, or money, or power in the light of the requirements of morality, or excellence, loses nothing, because from the standpoint of such an individual, any 'pay-off' from flouting such requirements cannot count as a genuine advantage. In other circumstances these sorts of things may be good and worth pursuing, but where one has absolutely no reason to pursue wealth or pleasure, these things fail to represent any sort of personal good or advantage.[4] Similarly, Watson holds that 'another way of illustrating the difference' between a man who thinks his sexual inclinations give him at least a prima-facie reason for sexual activity and one who thinks he has no reason to follow his sexual inclinations would be to say that 'for the one man, foregoing sexual relationships constitutes a *loss*, even if negligible compared with the gains of celibacy; whereas from the standpoint of the other person, no loss is sustained at all'.[5]

Watson and McDowell proceed to different uses of these parallel conclusions—Watson to a discussion of the role of (Platonically-understood) reasons and values in human freedom, McDowell to a defence of the Aristotelian assumption that virtue (the life of excellence) is in the best interests of the virtuous.[6] But for present purposes it is important merely to note how comfortably both of them rely on the existence of a tight connection between personal goods and reasons for

[4] 'The Role of *Eudaimonia* . . .', p. 365 f. For what appears to be similar views in Aristotle, see, e.g., *Nicomachean Ethics*, Book X, Ch. 3, 1173b 20–30; Book V, Ch. 1, 1129b 1–8.

[5] Op. cit., p. 210.

[6] McDowell and Watson appear to differ about the extent to which desires and emotions can exist independently of Reason and reasons, but this disagreement need not concern us in what follows. Also, McDowell subscribes to a version of the overridingness thesis and does not seem to allow for an admirable 'life of excellence' that was not a life of *moral* virtue. Cf. op. cit., pp. 368 ff. But I wish to leave room for such a possibility, and it is worth noting that our discussion of Gauguin's indifference to his own safety already illustrates how non-moral considerations of excellence may silence considerations of personal well-being.

action. Indeed, the assumption that such goods always give one at least prima-facie reasons for seeking them (or, contrapositively, that what one has no reason to seek is *eo ipso* not to one's advantage, not a personal good) is a fairly common one and would no doubt be widely viewed as trivial or obvious.[7] However, when this simple idea is combined in the above manner with the assumption that the requirements of virtue or excellence can silence all opposing considerations, it yields the conclusion that power or pleasure that must be forgone for reasons of excellence is no sort of loss to the virtuous individual and does not count among the good things, or advantages, he has, or will have, missed out on in his life. And this conclusion goes counter to the common belief that strict adherence to morality (or other ideals) can require the sacrifice of personal well-being or advantage. Clearly, something has to give, and it may perhaps be felt that what can give way with least cost to our deep-seated convictions is the assumption that opportunities for pleasure, wealth, and power offer the virtuous person no reason at all to flout the requirements of morality. In what follows, however, I would like to see whether we may not do better by continuing to accept this seeming insight, while at the same time relaxing the connection between goods and reasons that McDowell, Watson, and others have treated as axiomatic. But in order to do so we shall need to investigate some possibilities that McDowell and Watson virtually ignore.

II

Presumably, McDowell would not wish to deny that someone who feels no compunctions about injustice (or intemperance) can gain an advantage (something good for himself) through immoral actions, by making use of an opportunity, e.g., to steal money undetected or to win what would otherwise have

[7] Cf., for example, Joseph Raz's view (*Practical Reason and Norms*, London: Hutchinson, 1975) that a person has reason to promote whatever is in his interest (p. 34) and that values are reasons by definition (p. 25).

been a fair election by bribing local officials.[8] Such an admission does not prevent us from deploring (the actions of) that sort of individual, and it would be high-handed indeed to discount the immoralist's own sense that he gains something by his action. But charity then dictates similar respect for the point of view of the virtuous individual, and where pleasure or enrichment gives him no reason to do what is contrary to excellence, such a person will surely neither dwell on nor be struck by the advantages to be gained by immorality. For if he is thinking about the things he will miss and thinking of them *as* good, then the relevant considerations have presumably not been silenced, and he is not the virtuous individual we sought to imagine.[9] So charity seems to dictate that pleasure or wealth forgone (or ignored) for reasons of excellence are not forgone (or ignored) *goods* from the standpoint of the virtuous individual. And this seems a (further) vindication of the assumption that what we have no reason to choose is not (for us) a good thing. The point is not that virtuous individuals seek happiness, and do what is noble because they recognize that pleasure or wealth attained at the expense of the noble are not conducive to their happiness (well-being), for such eudaimonism would, once again, fail to respect the virtuous individual's own sense of what he is up to.[10] Rather, the idea is the precisely opposite one that certain things are not to the advantage of the virtuous, are no part of their happiness, because they are (independently) moved to do only what is noble or right. What is incompatible with nobility, with (moral) excellence, for that very reason does not appear to the virtuous individual as anything good.

[8] See McDowell, 'The Role of *Eudaimonia* . . .', p. 371 f.

[9] The truly virtuous does not chafe under moral requirements, and dwelling on advantages to be missed through virtue shows at the very least a certain ambivalence about such requirements. By the same token anyone who seeks and needs reassurance about whether morality pays shows himself to be the kind of person for whom advantages excluded by morality might (would) none the less carry weight and is not, therefore, fully virtuous. (This does not apply to those who treat the question whether virtue pays as a purely philosophical issue.)

[10] In 'The Role of *Eudaimonia* . . .' McDowell argues that we should not treat Aristotle as a (consistent) eudaimonist.

However, in saying that someone who chooses one of two alternative courses of action loses nothing (good) in doing so, one is speaking of what could or would have been and making some sort of counterfactual assertion.[11] The idea, for example, that the virtuous individual loses no personal good by not stealing money undetected or by not rigging an election seems to involve the claim that such a person loses no personal good that he would have gained by stealing or rigging the election. And this implies a particular view of the possible situation in which he would have stolen the money or rigged the election, according to which in that counterfactual situation stolen money or political power and success do not count for him as genuine advantages.

But if we do not wish to deny that the immoralist gains by stealing or rigging an election, we must surely have some reason for denying that the virtuous individual gains anything in the counterfactual situation where (in contradiction to his actual character) he acts similarly. And if there is a difference here in what is, or would be, gained, it is surely due to, or related to, some difference in the character of the two individuals and, most relevantly, to some difference in their character *in the two situations* in which they are respectively imagined to steal or do some other dishonourable thing. In that case, perhaps we can say that (part of) the relevant difference lies in the fact that a truly virtuous individual is the sort of person who, even if he acted out of actual character and stole money, would refuse to make use of things gained in this way. The latter counterfactual conceives of virtue as being deep-seated enough so as to have no tendency to disintegrate as a result of a single wrongdoing. But this idea seems plausible and corresponds very well to most traditional conceptions of virtue; and even though counterfactuals may constitute a general problem for philosophers, there is no reason perhaps to object to the counterfactual just mentioned, and I shall not do so.[12] In that case there certainly seems to be an important

[11] Cf. McDowell, ibid., p. 369 f.

[12] Cf. Aristotle's claim that the virtuous person does nothing contrary to

difference between the immoralist who has stolen and the virtuous individual in the counterfactual situation where he has stolen. What I doubt, however, is whether this difference can be of any help to McDowell. In fact, when we examine these cases more carefully, we see that they carry weight *against* the thesis about goods and reasons that is the cornerstone of McDowell's and Watson's arguments.

The (otherwise) virtuous individual who had stolen money would presumably feel remorse and try to return the money. But if the latter were impossible, he might still refuse to use it himself and prefer to give it all away to charity. Clearly, the desire to return the money can be plausibly accounted for in terms of a desire that others should not be unjustly deprived of what is theirs. But I think we can also assume that where the money cannot be returned, the desire to give it all to charity may not reflect any long-standing altruistic desire to give as much as possible to others, but result, instead, from a *refusal* to make use of the money for one's own purposes.[13] And in the circumstances such refusal makes sense only as a refusal to gain any advantage, or profit from one's own wrongdoing. But this precisely means that the virtuous individual sees his situation as one in which he will gain (keep) something good, beneficial to him, if he does not get rid of the money, as one in which the money represents some sort of personal advantage.[14] And if he is really virtuous

reason for the sake of bodily pleasure and would not feel pleasure (if he did act) contrary to reason. *Nicomachean Ethics* 1152a 1–4. Of course, we are speculating about what would happen if an individual acted against his own actual present and future character, but there is nothing inherently dubious about this. Even if a certain person is and always will be a football fan, we can speak meaningfully about what he would do if he failed to attend a certain important game (which he in fact will attend).

[13] I assume that virtue (as opposed to saintliness) does not require one to be such an altruist, but the argument could be reformulated so as to be free of that assumption.

[14] If he doesn't refuse to use the money and argues instead that it would be absurd for him to let his guilt stand in the way of using the money to help himself and others in worthwhile ways, then of course he is also viewing the money as a good thing. I assume that this won't occur in the present case, but later on we shall discuss a case where it actually seems *plausible* to suppose that a virtuous individual might retain an unjustly gained advantage.

and is fully resolved not to profit from his own wrong, the fact that he will or can thus profit by not getting rid of the money will not count with him as any sort of reason for retaining it. On the contrary, it is perhaps no exaggeration to say that in the situation where the otherwise virtuous man has done wrong to obtain the money, the fact that he would reap an advantage by retaining and using it is a positive reason for giving the money away, for *not* keeping it. So if before any act of stealing occurs, we can say of the virtuous man that he would gain nothing from stealing, this may in fact be true in virtue of another counterfactual that may be supposed to hold true in or of the (possible) later situation in which the stealing has already occurred and according to which our virtuous individual would refuse to use money he had stolen. And we can make sense of the truth of the latter —while respecting the point of view of the agent it concerns —only if we assume that in the later situation the possibility of profiting, of gaining something good, from a certain action is no reason whatever for choosing it. The thesis that personal goods are always reasons for acting to obtain such goods seems fundamentally damaged.

But if in certain counterfactual situations the virtuous individual would recognize unjustly acquired wealth or power as advantages he should not allow himself, it will be asked why he fails to do so in the actual situation where he acts in accordance with moral requirements. This question, however, arises from an ambiguity in the notion of 'not thinking of something as good' that I have so far refrained from mentioning. I said earlier that the virtuous man or woman would not dwell, e.g., on the advantages of stealing, but that does not mean that *if pressed*, they would deny that they were giving up something good, sacrificing their own well-being or advantage. It is one thing not to be thinking of something as good, another to be prepared to *deny* that it is good, and we need not regard the virtuous as having the latter tendency. Thus respect for the virtuous individual's standpoint does not, in fact, require us to deny the goodness (for him) of what he gives up, and there seems to be no inconsistency between virtuous individuals' actual and counterfactual attitudes towards ill-gotten wealth or power, only a difference of explicitness that is itself not difficult to account for. For where no wrong has been committed, the virtuous individual's focus on moral requirements can exclude all thought of the advantages of wrongdoing; but when he has acted wrongly and can gain (or has gained) an advantage thereby, he may well feel that it would be wrong for him to benefit from his wrongdoing, and then a focusing on moral requirements will precisely entail that he *be* thinking of wealth and power as (capable of) benefiting himself.

However, if the virtuous person may only implicitly, or *potentially*, recognize the advantages of flouting certain moral requirements, upholders of the view that reasons always accompany goods may perhaps have one last defence (a defence, however, that involves abandoning the relativization, or limitation, of human loss and advantage that McDowell argues for on the basis of that view). If attainable goods may properly fail to be reflected in an agent's awareness, why, it may be asked, cannot the same be true of reasons? In circumstances where a possible advantage can be gained by immorality, perhaps the advantage counts as a reason for flouting moral requirements that the virtuous individual simply has no business considering.

But such a move quickly entangles one in conceptual incoherence. If a reason for action is not properly reflected in the awareness of agents, then it has no place in practical reasoning or deliberation and is not action-guiding. And the idea of reasons for action that are not properly action-guiding, that are not *practical*, seems an outright contradiction.[15] The idea, however, of goods or personal advantages that have no claim on one's attention lacks this sort of

[15] By the same token, once one has been released from a promise, it would be absurd to suppose that one still had a prima-facie duty (reason) to keep that promise, just one that one had no reason to consider or be guided by. That is why it is so natural in such circumstances to speak, with Ross, of the abolishing of a duty.

absurdity. When one loses an advantage by acting one way rather than another, the fact of missed advantage is a matter of what would have been if one had acted otherwise, and it is just not obvious that agents always have reason to pay attention to such counterfactual possibilities. Duty may dictate a particular action in such a strong way that there is simply no reason to consider the possible advantageous results of acting otherwise.[16]

Moreover, the idea that certain goods or values are not properly reflected in the awareness of agents and so not practical has plausibility independently of the examples we have been discussing. Thus consider the value of spontaneity. If it is good to be spontaneous, spontaneity is none the less not something to be attained by following the practical principle 'act spontaneously'.[17] The good of spontaneity is not supposed to be reflected in the thought of the spontaneous-acting individual or to guide her in her spontaneous actions. And a similar point can be made about certain moral values. We are generally moved at a very deep level by the desire for (instinct of) self-preservation, and because we find it both understandable and morally acceptable that people should be thus moved, we think self-defence is generally permissible and legitimate. But what is thus permissible is that someone should defend himself because of a perceived danger and without any thought of morality. It would be absurd to expect the principle that self-defence is permissible (morally all right) to guide or even be considered by someone acting under threat of death, and we here have a moral value that corresponds to no immediately practical principle or reason for action.[18] Such examples of goods or values that are unsuited for conscious consideration should

[16] Cf. the economists' notion of satisficing.

[17] Cf. John Elster, *Ulysses and the Sirens*, Cambridge University Press, 1979, p. 168.

[18] For extended discussion and defence of the idea of non-practical values, and of the particular examples mentioned in the text, see my 'Morality Not a System of Imperatives', *American Philosophical Quarterly* 19, Oct. 1982, pp. 331–40.

make it easier for us to see how goods and reasons might go their separate ways when virtuous individuals are acting virtuously.[19]

III

However, quite apart from issues concerning the connection between goods and reasons, there are cases we have not yet considered that cast considerable doubt on the view that the virtuous lose nothing by being virtuous. McDowell makes the extremely general claim that no 'payoff' from flouting a requirement of excellence can represent an advantage for the virtuous person, and this seems very implausible when one takes into account the unforeseen and unforeseeable consequences wrong actions may have.

Imagine, for example, that your room-mate has tickets for a sell-out concert. You are yourself a budding violinist, are dying to see the concert, but cannot persuade your room-mate to part with his ticket. Of course, you wouldn't dream of using dishonest means of getting the ticket from him, but what if you did? What if you stole the ticket or, knowing that he was careless, followed your room-mate about until he mislaid, dropped, or otherwise lost the ticket and you could pocket it for your own use? It is possible that having done all this, you should end up sitting next to Isaac Stern, that you should strike up a conversation with him, say something to impress him, and begin a friendship that helped to advance your career as a violinist in ways you could never have managed on your own. And all of this would be due to unforeseen circumstances, to luck.

Naturally enough, a virtuous person would want to make

[19] In Chapter 1, it was left open whether an individual might have no reason to pursue or prepare for a future good that lay beyond his present desires and interests. If Bernard Williams is correct in believing this to be possible, then we have yet another instance of goods that fail to give rise to reasons, but Williams's examples, if valid, are ones where our present desires, hence reasons, fail to reach out to and include some later personal good, whereas in the cases we have been considering, moral requirements positively *exclude* some present or future advantage.

amends for his wrongdoing. But since your room-mate is no musician, you may not be able to make it up to him by introducing him to Stern. (The advantage gained from knowing Stern may be assumed to be untransferrable.) But since the harm we are imagining you to have done to your friend is relatively small, making it up to him may not be the most difficult thing in the world or take the rest of your life. (I leave it open whether this would require confessing what you have done.) And given the disproportionately great benefits you reap, the total difference friendship with Stern can make to your life, it may be easier to make amends if you maintain your friendship with Stern rather than abandoning it.

However, apart from the felt need to make reparation, there is the intrinsic shamefulness of your actions, and it may be said that it would be wrong of you to profit from them for the same sorts of reasons that made such an attitude appropriate in the example(s) mentioned earlier. But the present case is importantly different. When one's motive (intention) is to gain power or wealth, then the relinquishing of that power or wealth seems an appropriate act of expiation. The moral character of an act is determined by its motives (intentions) and such expiation seems relevant to the wrong committed (the self-punishment fits the crime). But when someone has inadvertently gained by doing something wrong, what is thus gained is not so immediately relevant to expiation, and given, further, the sheer disproportionality between the wrong committed and the great advantage gained, your friendship with Stern need not perhaps be a pawn to your desire for atonement. Perhaps it would be more appropriate for you to use your own subsequent advantages as a musician for the benefit, in part, of less fortunate musicians, rather than turning your back on this opportunity to help yourself and others in what might well seem a gesture of morbid moral fastidiousness rather than of moral courage. (Given the disproportionality, wouldn't it also be morbid to *keep* the connection with Stern and its attendant advantages but spend *the whole rest of one's life* expiating a sense of guilt?)

But also because of the disproportionality in the present case, it seems harder to deny that you would gain something if you were to steal the ticket. The advantage to be gained is writ so large and combined in such a way with the fact of inadvertency, of luck, that it seems even a virtuous individual would recognize the friendship with Stern as having proved advantageous and not flinch (well, he might just flinch) from making use of such an advantage. And in that case it would be preposterous to deny that you miss out on a (possible) advantage when, in circumstances where acquaintances with Isaac Stern would in fact be forthcoming, you virtuously and, of course, all unthinkingly do not steal your room-mate's concert ticket. Independently of any argument for the disconnection of reasons from goods, the present example is a challenge to McDowell's presumptively Aristotelian thesis that the virtuous individual loses nothing (good for him) in acting virtuously.[20]

IV

We have accepted Watson and McDowell's claim, inspired or animated, as it is, by a certain reading of Plato and of Aristotle, that our values, our standards of (moral) excellence, can deprive certain facts of their accustomed status as reasons for action. But we have done so while maintaining that one need not accept their further assumption that reasons automatically attach to goods or their Platonic/Aristotelian conclusion that one who acts in the light of certain standards

[20] But the example we have used may be no better than ones that Aristotle himself was acquainted with and apparently undaunted by. For both he and McDowell are aware that virtue can require one to sacrifice one's life, or health, on the battlefield, and it seems equally preposterous to deny that the virtuous soldier can lose in personal well-being. Perhaps Aristotle or McDowell has a way of drawing the sting from such examples that we are just not thinking of, but, on the other hand, consider Aristotle's wildly hedonistic remarks, in discussing such cases, about the virtuous soldier's preferring the intense pleasure to be gained from a year's nobility and virtue over the blander pleasures of a longer but more humdrum existence. (*Nicomachean Ethics*, 1169a 2–25.)

or values may be seen as incurring no loss in doing so. There is, in fact, some precedent to this willingness to go only half the way with Plato and Aristotle, and it will not perhaps surprise the reader if in this connection I mention the name of Kant. What may, however, be surprising is the fact that Kant invokes the same notion of silence that McDowell appeals to, when he (Kant) describes the virtuous person's attitude towards considerations of personal well-being that might oppose the fulfilment of duty. Near the close of *The Metaphysical Principles of Virtue*, Kant presents a catechism of questions and answers and puts into the mouth of the well-instructed student the words: 'I should not lie, though the advantage to me and my friend be as great as ever you please. Lying is mean and makes a man unworthy to be happy. Here is an unconditional constraint by a command (or prohibition) of reason, which I must obey. In the face of this, all my inclinations must be silent.'[21]

Kant's idea that what goes against one's sense of right can be silent, even though an advantage could be gained by acting wrongly, embodies McDowell's claim that one has no reason to act against virtue (the right), but denies the inference that what thus offers no reason and is silent does not constitute a (missed) advantage. But is Kant being consistent here? In the *Critique of Practical Reason*,[22] he speaks of the priority of the moral law to the idea of the good, and does he not mean that what offends the moral law is not a good thing? Indeed he does; but earlier, in the same chapter of the second *Critique*, Kant points out an ambiguity in the term 'good' (*bonum*) and distinguishes between the good (*das Gute*) and well-being (*das Wohl*).[23] He then makes it clear that he is willing to grant that acting against the moral law may be to our advantage and means to deny only that such a thing can

[21] *The Metaphysical Principles of Virtue*, Indianapolis: Bobbs-Merrill, 1964, p. 150. This is part two of *The Metaphysics of Morals* (a different work from *The Foundations of the Metaphysics of Morals*).

[22] *Critique of Practical Reason*, Chicago: University of Chicago Press, 1949, Part I, Book I, Ch. 2, pp. 171 ff.

[23] Ibid., pp. 168 f.

be good in itself, something a disinterested party would call good. Thus (to recall Chapter 3) when Kant says that happiness without virtue is not a good thing, he is only saying that it is not good from a disinterested standpoint. He is not denying that the immoral person may be *well-off* or claiming that good in the sense of well-being is defined or constrained by the moral law, by the right. Indeed Kant seems to think of happiness and well-being in a very conventional way, and though, of course, he argues that happiness and virtue do coincide in the long run, this is effected via various postulates of practical reason having to do with God and the immortality of the soul, and there is never any suggestion that morality requires no sacrifice of personal well-being in *this* life.

At any rate, apart from considerations of an afterlife, the Kantian view of things seems closer to our deepest understanding of morality and ourselves than anything derivable from Plato or Aristotle along the lines suggested by McDowell (or Watson). We seem to have the capacity for shaping and governing our lives according to certain ideals or standards, and in the light of these, our own very real advantage often appears to carry no weight whatever in our practical deliberations and decisions. [As Kant puts it, all actions may be directed at some (or the) good, but this good need not be, or even be consistent with the agent's well-being.] [24] Of course, in taking this view of things, I have, like Kant, taken a fairly conventional view of the nature and contents of individual well-being, but this too may be necessary in order to give expression to our sense of the *relation* between moral ideals and personal advantage.

At this point, however, it is also worth noting that the Aristotelian idea of eudaimonia, of living or doing well, is subject to an ambiguity very similar to that noticed by Kant in the idea of the good. Living well (living a good life) may mean either living a life of great personal well-being or living admirably, and the lives we most admire, or think well of,

[24] Ibid.

are not necessarily the lives we think happiest or most filled with personal advantages. Of course we may recognize the ambiguity in 'living well'—Aristotle may not, but McDowell and others certainly do—[25] while still maintaining that the virtuous or admirable lose no personal advantages in following their ideals unhesitatingly. But the important thing here is not to be led into a premature and unjustified acceptance of this latter assumption through confusion or conflation of the two meanings of 'living well'.

A similar, and potentially more misleading, ambiguity attaches to the notion of value. A soldier who gives no heed to his own safety may be said to be placing no value (or a low value) on his life (his health, his safety).[26] But even if he can lose nothing he values in those circumstances, that offers no support to McDowell's belief that he risks no personal loss, no loss of well-being. In this context the term 'value' does not, I believe, refer to personal well-being, and in saying that the soldier doesn't value his life, health, or safety—that in the circumstances these things lack value for him—we are not saying that these things are not elements of his well-being. For we could as easily have said that such a soldier places no value on *his own well-being*. In claiming that he doesn't value his safety, etc., I think we are saying, rather, that from his distinctive standpoint, considerations of health or safety give him no *reason* to avoid the risks he is taking. On the other hand, it is easy to see that a phrase like 'has no value for him' may *sometimes* indicate that something makes or would make no contribution to an individual's well-being, and a confusion of these two meanings may be part of what tempts one to view the intrepid soldier as a confirmation of McDowell's thesis.

[25] Cf. McDowell, 'The Role of *Eudaimonia*', p. 368; Kathleen Wilkes, 'The Good Man and the Good for Man' in Rorty, ed., op. cit., p. 343.

[26] A fact noted, for example, by Ronald Dworkin in 'What is Equality? Part I: Equality of Welfare', *Philosophy and Public Affairs* 10, 1981, p. 210 f.

V

So far we have largely stayed clear of examples concerning pleasure. We have defended the idea that wealth, power, and success are personal goods even when obtainable only by unjust means, and have resisted any relativization of these goods to those individual circumstances where a person has reason to pursue them. But the view that pleasure has a similarly univocal character as a personal good is fraught with special difficulties, and it is to these that we must now proceed.

McDowell asserts his claim that a failure to realize a pleasure inconsistent with virtue (excellence) is no loss to the virtuous individual, with particular reference to those pleasures which would be consistent with virtue and count as personal goods in different circumstances, but which given agents at given times cannot achieve except at the cost of their ideals. And I believe we can cast doubt on his treatment of such cases by essentially the same arguments used above. Even if the virtuous man has no reason to pursue a certain pleasure, that fact need not convince us that the pleasure would not, if pursued and attained, count among the (momentary) goods of his life. (Certainly, the pleasure might have bad consequences, but that is irrelevant here. Someone who says pleasure is always a good thing is not denying that some pleasures have bad consequences.)

On the other hand, we might, like Aristotle,[27] deny that the virtuous individual would get pleasure from flouting moral requirements. Even if such an individual forgoes what *others* would enjoy, we might claim that he loses nothing that he himself would enjoy. Interestingly enough, however, McDowell does not take this view of pleasure excluded by virtue. He seems to assume that the virtuous person would (or might) enjoy a certain pleasure but that he or she loses nothing by not pursuing it in virtue of the absence of any reason to pursue it. Thus Aristotle and McDowell seem

[27] At 1152a 1–4, as I pointed out in footnote 12 above.

agreed that failure to pursue a pleasure inconsistent with virtue involves no loss to the virtuous individual, but their reasons are different.[28] The one says (roughly) that what was not pursued would not have been pleasurable, the other that any pleasure ignobly attained would not have been a personal good for the virtuous individual. Let us consider these alternatives.

Although Aristotle simply denies that the virtuous man would feel pleasure contrary to reason (virtue), some recent writers have made more specific suggestions. David Pears has advanced the view that the virtuous individual would feel revolted in acting contrary to reason, and Watson seems to believe that such a person might feel too anxious to enjoy a pleasure contrary to his own sense of values.[29] But are these the only possibilities? Mightn't the (otherwise) virtuous individual who acted contrary to reason or virtue simply not let him- or herself enjoy what he/she was doing: in cases of adultery, for example, by turning off emotionally, making oneself frigid, or deliberately distracting one's attention from what was going on? And isn't such unwillingness to enjoy, best construed, as before, in terms of a *refusal to derive any good* from the wrong one is doing? (Perhaps such a person is also in a primitive way trying to undo an ignoble act by subverting its normal consequences.) The idea that virtue can prevent someone from feeling pleasure in certain things may support the view that the virtuous lose nothing by being temperate and just about pleasure, but clearly it gives us no reason to deny that (existent) pleasures are always personal goods.

McDowell, however, seems committed to just such a denial and we must now ask how his case might at this point proceed. Perhaps, for example, the very fact that a virtuous individual may later have no regrets whatever about missing

[28] They also agree that one may get pleasure from renouncing such pleasure, and nothing I shall say or have said is meant to be inconsistent with holding that the life of virtue has its own distinctive satisfactions (pleasures).

[29] See Pears's 'Aristotle's Analysis of Courage', *Midwest Studies in Philosophy* III, 1978, p. 276, and Watson, op. cit., p. 211.

a pleasure he would actually have enjoyed, shows that the missed pleasure, had it occurred, would not have been an even momentary personal good for him. But in the light of what we have already said about the partial disconnection between good and reasons, the present argument should be unpersuasive. For the absence of regret after the fact may simply reflect the earlier sense of having *no reason* to pursue that pleasure. Or, rather, the two attitudes may be, respectively, the natural before-the-fact and after-the-fact expressions of the same underlying disposition to virtue. Anyone who does have regrets about his own failure to seek and get a pleasure contrary to virtue thus shows himself to lack a full and unambivalent commitment to (his own conception of) virtue. And if the absence of regret naturally corresponds to the absence of reason in this way, then the former need not be any sort of indication that what has been missed is not a good thing.

However, we have not yet considered the most plausible sort of counter-example to the thesis that all pleasures are personal goods. The pleasures of sadism and heroin addiction, for example, are not merely sometimes inconsistent with excellence or virtue, but are thought of as inherently and universally shameful, degrading and/or wrong; and it is particularly difficult to maintain any sense that such pleasures are none the less personal goods. But although we would not ordinarily reckon such pleasures among the (momentary) goods of life (or of some lives), I suspect that our reluctance here may largely rest on mistakes and confusions of a sort we have already dealt with.

For example it would be senseless at this point to argue that sadistic pleasures are not personal goods because no disinterested being would wish them to exist, because it is not intrinsically good that they should exist. A disinterested being might also disapprove of a vicious man's being happy, but it hardly follows that the vicious man is not well off. And we have no more reason on that score to deny that sadistic pleasures are personal goods.[30] (I am not *claiming* that it is not intrinsically good for someone to enjoy sadistic pleasures or for a vicious man to be happy, only pointing out, rather, that such judgements of intrinsic value do not automatically yield conclusions about personal well-being. I have no very well defined intuitions about intrinsic goods and evils and I am deliberately restricting myself to claims about what is personally good and evil.)

By the same token, even if most of us have no regrets about never having experienced the pleasure of a heroin 'rush', even if the reformed (or unwilling) sadist may have no regrets about missing the chance to get some sadistic pleasure, it simply doesn't follow that those pleasures wouldn't have been personal goods, for we have already seen other cases where possible personal goods are neither reasons for action nor potential subjects of regret. (The reformed sadist may also feel revulsion at his former doings, but so too may the penitent coward or adulterer.)

Why, then, is it so much more difficult to regard sadistic or addictive pleasures as good(s) than to take that view of pleasures that are only sometimes inconsistent with (most people's ideas of) excellence or moral requirements? The *essential* shamefulness of the former may itself perhaps be a factor in this difference of attitude, but I suspect that the unfamiliarity of sadism and drug addiction may also play a role. Most of us lack a taste or even a capacity for full-blown sadism,[31] and as a result don't really understand how people can get pleasure in that way. This somehow makes us

[30] They may not count as benefits, but neither do most pleasures. Joel Feinberg has pointed out that pains may not be harms (see his 'Legal Moralism and Freefloating Evils', *Pacific Philosophical Quarterly* 61, 1980, pp. 122–55), but this does not mean they are not momentary evils in our lives. On the distinction between the personal and the intrinsic goodness of sadistic enjoyment, see A. K. Sen's 'Utilitarianism and Welfarism', *Journal of Philosophy* 76, 1979, pp. 463–89.

[31] To the extent that a virtuous person or anyone else is incapable of sadistic satisfactions, he or she may be said to lose nothing by failing to act sadistically and feel the pleasure that another might obtain in similar circumstances.

sceptical, I think, about whether the sadist really does get pleasure, anything *we* would call pleasure, from his activities. In the same way, the unimaginability for us of what the heroin rush is like allows us scope for doubting whether the addict is on to anything even momentarily good. It makes us wonder, for example, whether he may not physically need the rush rather than actually enjoy it or do the whole thing simply *pour épater les bourgeois*.

In that case, the goodness of sadistic and addictive enjoyments may be obscured by a partial but perhaps inevitable other-minds problem. It may be our own limitations (together with certain confusions) that make it difficult for us to acknowledge the goodness of what sadists and addicts enjoy. And one may well wonder whether there really are any good reasons to deny the thesis that pleasure is always at least a momentary good.[32] Like wealth, position, power, and success in a career, the status of pleasure as a personal good may be independent both of one's particular circumstances and of one's deepest ethical convictions.

[32] Of course, some pleasures are 'mixed' with pain or unpleasantness, and there are such things as the 'false' pleasures enjoyed when one has false hopes concerning the future. But even these may be good to the extent that they are pleasurable. Their pleasurable aspect, at least, seems to represent some sort of personal good. I have throughout the preceding discussion been trying to avoid discussing the ontology of pleasure, but I should at least say that I am not assuming that all pleasures are in the mind. Pleasures may essentially involve objects in the world, and when one speaks, for example, of the 'pleasures of the table', one may not be referring simply to inner sensations of pleasure, but rather to total experiences (note the word) of deriving enjoyment from *eating food*.

6

Stoicism and the Limits of Human Good

We have just been criticizing attempts to set ethical limits to the goodness of wealth, success, power, and (worldly) pleasure. And earlier, in Chapter 2, it was maintained that love and the affection of friends were human goods despite their unsuitability for inclusion among the goals of rational life-planning. Now, however, we must consider some quite different ethical ideals, embodied in Stoicism, that can lead one to question whether *any* of the above things is truly good. (I shall for the most part treat Stoicism as an ideal type and smooth over various differences among the Stoics.)

I

The Stoics praised and recommended a form of self-sufficiency, *autarkeia*, in which men would transcend normal human limitations and attain a higher (and more godlike) form of existence.[1] The ideal Stoic sage, or *autarkes*, is supposed to need and want nothing outside the immediate control of his own will and reason; his happiness consists in the exercise of reason, in virtue, rather than in any goods whose enjoyment or attainment depends on external circumstances or luck. (Notice the assumption that the exercise of rationality and virtue, once these are attained, is entirely

[1] It is surely not accidental that the life Stoicism recommends resembles in so many ways the sort of existence God or a god has been traditionally imagined to have. Of course, the Stoic sage, unlike a god, could be killed or harmed, but in his indifference to those possibilities the sage acts, in effect, as if he were invulnerable to death, harm, and pain, as if he were some sort of higher being other than a man. Compare Plato's *Republic*, Book VI (S. 500): '[the philosopher] contemplates a world of unchanging and harmonious order, where reason governs and nothing can do or suffer wrong; and, like one who imitates an admired companion, he cannot fail to fashion himself in its likeness.'

within our control.[2]) For that reason, the sage has to be free of all attachment to worldly goods and pleasures—he is allowed, of course, the rational pleasures of virtue—and in no way dependent on other people. He is also required to be indifferent to pain and death, and thus to all threats of tyrants and the unjust.

Stoicism treats the above ideal of *autarkeia* as the standard against which all goodness must be measured. Wealth, success, and power (even life itself) thus turn out not to be good things, even for humans, because the desire for such things—and the fear of their loss—are incompatible with self-sufficiency and make one's happiness a prey to external circumstances; and neither can love or affection be any sort of personal good, since it makes one's happiness dependent on the (occasional) presence and/or happiness of other people. In that case, common human opinion about the good things in life is, according to Stoicism, largely mistaken, and most of human life is filled with the irrational pursuit of things that are in no way good.[3] The only hope is to cast off these illusions and achieve the virtuous rational detachment in which alone human good is actually to be found. But this requires us to free ourselves from all worldly desires and emotional dependencies. And there was in fact some disagreement among the Stoics about how independent desires and emotions were of reason and the will, and in particular of the (rational) will to get rid of them. Some, like Chrysippus, seem to have held that it was in a person's power to stop desiring and feeling irrationally as soon as he had perceived the validity of Stoic doctrines. Others, like Posidonius, assumed that the emotions had a recalcitrant life of their

[2] Joel Feinberg ('Problematic Responsibility in Law and Morals', in *Doing and Deserving*: Princeton University Press, 1970, pp. 33 ff.) points out that a man's thought processes—and why not also exercises of virtue?—can be interrupted by such things as sneezing fits, which are entirely beyond his control.

[3] There is no reason in the present context to object to the slide from the non-goodness of things to the irrationality of pursuing them. Moral constraints that might make it rational to pursue what was not a personal good are not here at issue.

own and that it could take time to subdue them.[4] But despite these differences, there was fairly general agreement among the Stoics that freedom from desire and emotion[5] was worth aiming for and represented the best sort of life for man.

II

These recommendations make sense, however, only if men are capable of living a life of self-sufficiency or at least better off for attempting to do so. And how plausible is it to assume that they are? Consider love and friendship. Another, higher sort of being that never went through the dependency of prolonged childhood might never need or want friendship or love. But we are not beings of that sort: we seek love and affection from the start and never really outgrow that quest or its urgency; and in the process of establishing an identity apart from our parents or elders, we rely on close relationships with others of approximately our own age. (Perhaps these facts, this way of seeing things, was simply not available in the ancient world.)

So in the abstract, freedom from dependence on others may perhaps be an ideal or admirable thing, a perfection. And since (the need for) love and friendship are forms of such dependency, they may well be considered weaknesses. [Epictetus claims that one can love people without emotionally depending upon them, or their well-being.[6] But this involves, I believe, the conceptual error of supposing that one can have an emotional attitude like love and yet remain free to pick and choose among one's emotional reactions, and so

[4] Cf. F. H. Sandbach, *The Stoics*, London: Chatto and Windus, 1975, pp. 59–67.

[5] Emotion at least in the sense of something one is subject to, not entirely (rationally) active with regard to, emotion in the sense of passive or passionate emotion. The Stoics sometimes referred to non-passive emotions (the delight in virtue, for example), but the term '*pathe*' was used whenever the passive and objectionable kind of emotion was in question, and our own term 'emotion' entails, I think, a similar element of passivity, of being subject to and not entirely in control of something.

[6] See his *Discourses*, the section entitled 'That we should not be affected by things not in our own power'.

be able to (decide to) be utterly calm and undisturbed, e.g., at harm done to a loved one. Where we are not subject to, (partially) passive in the face of, certain feelings, our attitude is perhaps simply not describable as love.] But if love and friendship are weaknesses, they are *basic* human weaknesses: by which I mean that they are weaknesses so endemic to our nature that if one seeks, as the Stoics urge, to avoid being subject to them, one is likely to get oneself into a worse position than one would be in if one simply accepted the weakness in oneself. The tendency towards, the need for, the various affections of love and friendship may be basic weaknesses in this sense because if one attempts to be utterly free of them, one will simply cover up one's needs and feelings and in the process give them free rein for subterranean mischief and eventual destructive effect within one's life. Thus the Stoic ideal of emotional detachment seems to be an illusion for us, an admirable ideal perhaps, but one that we are probably just not capable of. And if we make this assumption, then the injunction to cultivate detachment to the greatest extent possible will seem highly problematic, and the Stoic who claims to have achieved emotional detachment will be thought to be papering over his or her own deep, possibly thwarted, yearnings for love and affection.

But where does this leave us with the question whether love and friendship are (personal) goods? Our discussion has been assuming that *autarkeia* is some sort of perfection and that love and friendship (at least emotion- and affection-involving friendship) are not goods for an ideal being. And since the standard of personal good might plausibly be held to be what is good for such a being, we may at this point be tempted to conclude that love and (affectionate) friendship, as well as wealth, power, and success, are simply not good for us humans—however inevitably we may desire or pursue them.

However, in earlier chapters, we considered a number of forms of relationality, or relativity, that inhabit various kinds of good and virtue; and the idea that friendship and love, even if not good for every possible being or person, may none the less be good *for men and women*, personal goods relative to *our sort* of nature, now fairly cries out for consideration. If the ideal of self-sufficiency sets the standard for personal good and we granted the desirability and feasibility for us of the goal of indifference to love and affection, then we might have to allow that those things are no sort of human goods. But if, as I assume, we inevitably need some kind of love and friendship and stand to lose from attempts to avoid them altogether, then they may be good for the likes of us, even if they are not good for every being imaginable. The most, then, that the Stoic can show, is that love and friendship are not *universal* personal goods.

And similar reasoning no doubt could be used to argue that success, (worldly) pleasure, and wealth (or material well-being) are also good things for human beings, even if no ideal being would need or desire them. Moreover, since what is (a personal) good for humans is supposedly a function of basic human needs (desires), such mankind-relativity of goods (like the forms of relativity described in Chapters 1 and 2, above) runs no risk of the sort of relativism which, by making values depend on choice or belief, makes them seem subjective. What is not absolute may none the less be objective.

III

Furthermore, the ideal of *autarkeia* can function for us as the measure of personal good only if we are willing to concede that the life of such a being would inevitably be better than our own. For if a human being might—by having the good luck never to encounter any major misfortune pertaining to love, friendship, success or material well-being—be better off, moment for moment, epoch for epoch, than a self-sufficient being, then love, friendship, etc., may count as absolute goods, just very risky and unstable ones. If, however, we grant that a self-sufficient being would have an unconditional preference for being *immune* to the reverses and losses that so often accompany wealth, love, and friendship, we may perhaps conclude that his life, his existence, *is* essentially better than our own, and his indifference to anything that involves risk will then make love, wealth, and success appear lacking in personal value.

And this point of view can be reinforced by the further consideration that a self-sufficient being might prefer to live (e.g.) without love, not only because of its inherent riskiness, but also because love is a good thing relative to needs he deems it better not to have at all.[7] Thus consider the attitudes a self-sufficient being is (believed) likely to have towards the purely physical appetites. While perfectly willing to grant that these appetites are not phenomenologically like itches or pains, which no one could reasonably wish for in order to have the pleasure of relieving them, he may still find copulation or the ingestion of food intrinsically undignified and repellent. This evaluation would naturally create an aversion to (real) satisfactions gained by these means quite apart from any attendant risks of frustration and disappointment, and something similar can, I think, be said about love; for a self-sufficient higher being might well regard dependency on others, with its accompanying emotions and satisfactions, as inherently undignified, absurd, perhaps even grotesque. (Here we may have yet another ethical other-minds problem.)

But to treat such a higher being as the measure of value, we must in some way share his evaluations. We must, for example, accept his total aversion to risk as a rational ideal, and I wonder how many of us would really be willing to do so. And I wonder too whether we will agree with such a being that there is something absurd or undignified about copulation and ingestion.[8] Surely, any feeling on our part that there is something ludicrous or repugnant about these things may be interpreted as a sign of self-disgust, of a puritanism that

[7] Here my discussion is much indebted to ideas, about (the value of) the appetites, to be found in Gary Watson's 'Free Agency', *Journal of Philosophy* LXXII, 1975, pp. 205–20.

[8] W. Somerset Maugham once said of sex: 'the pleasure is momentary, the expense damnable, and the posture ridiculous'.

finds the body degrading, and if the latter attitudes seem suspect, we need not take the attitudes and preferences of an appetiteless higher being as definitive of what is good. Love and sexual or gustatory enjoyment may then count as good things that such a being simply lacks the capacity and inclination for.

Furthermore, when we stop wondering what such a being would think of us and our ways and consider what we, upon reflection, may want to say about *him*, our sense of the goodness of love and appetitive satisfactions may receive further reinforcement. There may be some personal goods that a self-sufficient being can enjoy but we cannot, and perhaps a being freed from our appetites and from various human vulnerabilities and limitations can be thought of as higher, more perfect, more admirable as a type, than we. But we have for a long time now been stressing the difference between ideals of admirability, excellence, or virtue and the sorts of good things that make for individual well-being (or happiness), and when we actually think about what it must be like to be such a higher being and about what his sort of happiness would consist in, we may well wonder whether he is ideally well off or even better off than (some or most) human beings.[9] With love, friendship, the appetites, and all striving for achievement absent from his existence (and he is really not even a he or a she on these assumptions), isn't that existence apt to strike one—and to continue to strike one—as lonely, somewhat empty, and/or not very interesting? Perhaps (*pace* the Stoics) we should not want such an existence even if we were capable of having it; perhaps it corresponds to nothing that we (or most of us) can acknowledge and see as a form of happiness. (Again, our recurrent other-minds problem.) And in that case, it will be tempting to say that the life of a higher being of the sort imagined, however, free from human weaknesses and limitations, would

[9] I am indebted here to Thomas Nagel's way of putting things in 'What Is It Like to Be a Bat?', reprinted in *Mortal Questions*, Cambridge University Press, 1979, pp. 165–80.

be lacking in satisfaction for the being in question. That would be yet another reason for thinking that the things we enjoy, but such a being cannot, are personal goods. (However, for reasons to be made clear shortly, this need not mean that the invulnerability of a powerful and appetiteless being cannot also count as a personal good.)

None the less, there may be important differences among some of these goods, and someone willing to acknowledge the value of pleasure, friendship, success, love, and wealth might none the less insist that the last of these, wealth, was a merely instrumental good and not, like the others, a personal good quite apart from its consequences. (I here ignore the sort of hedonism that treats success of various kinds, and even love and friendship, as merely instrumental goods.)

Now this view is perhaps most plausible when wealth, or the possession of a certain amount of money, is distinguished from material well-being or comfort; for the best evidence for the merely instrumental character of wealth comes from cases of misers or recluses where wealth lacks its usual consequences and appears to lack all value for its possessor. But in such cases wealth also fails to serve material comfort and the latter's status as a personal non-instrumental good is thus in no way impugned. However, it is also not clear to me that even wealth has to be regarded as a strictly instrumental good. Some people who live in large cities do not take advantage of their cultural opportunities, but say that they none the less like having such a *choice* of films, concerts, and the like. The opportunity is valued for its own sake, and such people often consider themselves fortunate by comparison with out-of-towners who manage, in their short visits, to take in many more cultural events. (The short-term visitors may feel the same way.) But, then, in the light of such examples, wealth too may seem a kind of valuable opportunity quite apart from its actual *use*. A miser may be pitiable as the victim of a pathological *inability* to spend money, and people who live in large cities might well do *better* to take more advantage of the opportunities they take for granted, but that need not mean that wealth and the availability of culture are of *no* value in themselves. The above list of (possibly absolute) personal goods may not be quite so heterogeneous as one might think.[10]

IV

In addition, some possible goods do not appear on that list and have not yet been mentioned, and I would like to close this chapter by briefly considering what is perhaps the most important among these.

Earlier, in Chapter 1, I pointed out that old people often envy the young for having so much of their lives remaining and that youth returns the compliment by pitying the aged for having so little time left. These attitudes appear to be independent of any assumption about how (well) young people are likely to fill, or fail to fill, their lives and thus seem to express a view of life (or perhaps of life *compos mentis*, since no one envies the comatose, or living vegetables, for the length of their sort of existence) as a non-instrumental personal good for its possessors. This stands in marked contrast with the widespread view that it is only life's 'contents'—the pleasures, attainments, virtues, or whatever that life

[10] In speaking of what may be of (no) value in itself, I am not here invoking the notion of intrinsic goodness mentioned in Chapter 5. I am restricting the discussion to personal goods, rather than considering what may or may not be thought good from some impersonal or disinterested standpoint. But in contrasting personal goods that are merely instrumental with others that are not, it is tempting and natural to speak of the latter as (personally) good *in themselves* or *intrinsically*, and this can lead to confusion. (I allowed myself to speak this way in Chapter 3 because the distinction between personal goods and things from a disinterested standpoint was not there an issue). The ease with which the term 'intrinsic' is used in contrast both with the instrumental and with the personal has, I think, been partly responsible for the mistaken opinion that what is not good from a disinterested standpoint is automatically not a non-instrumental personal good.

makes possible, rather than (conscious) life itself—that are really good, a view that I certainly cannot refute, but which age's envy and youth's pity seem to belie.[11]

What I am suggesting by way of an alternative is that we take the latter attitudes at face value and think of conscious life itself as a personal good. As such it would also be the necessary condition of all other personal goods, but if opportunities qualify as genuine personal goods, conscious life as the 'space' in which other goods occur may perhaps also count as such. Just as personal freedom and political liberty possess a value independently of how well or wisely they are used, life itself, life lived in possession of one's faculties, can be a good thing for its possessor even when filled with tribulations and unpleasantness. And the tenacity with which we persist in staying alive may indicate, not an instinct that makes us irrationally cling even to an unpleasant or painful existence, but a fundamental acknowledgement of the fundamental good of being alive.

All this is hardly original. Aristotle asserts the non-instrumental goodness of being alive and of knowing and seeing things, and David Wiggins has recently emphasized the possibility of treating the basic desire for self-preservation as something other than irrational.[12] But the attitude of the young and old towards one another offer what is perhaps a new kind of evidence against the view that the good of life depends solely on how life is used or filled, and our treatment of wealth (and personal and political freedom) not only helps us to see how a space for other goods may itself be a personal good quite apart from how it is used, but

[11] The view in question is held by hedonists and many others: e.g., by G. E. Moore (*Principia Ethica*, Cambridge: 1959, esp. pp. 156 f.); by Bernard Williams (in 'Persons, Character and Morality', in A. Rorty, ed., *The Identities of Persons*, Berkeley: University of California Press, 1976, esp. pp. 207 ff.; and in 'The Makropulos Case: Reflections on the Tedium of Immortality', in his *Problems of the Self*, Cambridge University Press, 1973, pp. 86 f.); and by Harry Silverstein ('The Evil of Death', *Journal of Philosophy* LXXVII, 1980, pp. 401–24).

[12] e.g., in the *Nicomachean Ethics* Book IX, Ch. 9, 1170a 15–1170b 19. Wiggins in 'The Concern to Survive', *Midwest Studies in Philosophy* vol. IV (*Studies in Metaphysics*), p. 420.

opens up the distinct possibility of a whole hierarchy of such personal goods related as space to 'filler', or as form to content (matter). Conscious life may be a formal good which makes possible the possession of personal freedom; this 'content' may then itself be a formal good, or space, in which wealth can be attained, and the latter good in turn may constitute a yet lower-order space in which lowest-order non-formal (and in that sense 'material') goods like pleasure are facilitated.

Such a hierarchy may not, however, be rigid, since (for example) in some non-democratic societies, and to some extent in our own, it is wealth that makes personal liberties possible, rather than the reverse.[13] But in any case the idea of such a shifting hierarchy points up a rather neglected, further way in which goods (perhaps even virtues) may be related, and must surely take its place alongside the earlier notions of time-preference, relativity, and dependency in attempts to explore the complexity of human good.[14]

[13] Compare Rawls on the 'worth of liberty', in *A Theory of Justice*, Cambridge, Mass.: Harvard, 1971, p. 204.

[14] Compare and contrast Aristotle's interlocking metaphysical hierarchy of forms and matters.

Conclusion

In these chapters, I have not had much to say about the cardinal virtues of wisdom, courage, temperance, and justice, but have tended, instead, to rely on certain recent treatments of these virtues. I am afraid that this indebtedness is only incompletely recorded within the above text and in footnotes. But if we are to make progress in understanding the principal virtues, I think we would do well to make use of recent work in this field, and my own discussion of relative and dependent virtues is intended to show only—but I believe it is an important point—that the subject of virtue as a whole requires qualifications and complications that treatments exclusively of the principal or cardinal virtues tend to ignore.

I have also defended the idea of admirable personal traits, or virtues, that run contrary to morality and have done so within a framework that presupposes the objectivity of right and wrong. But it is worth noting that the assumption of moral objectivity need not force one into over-confidence about the attainment of moral knowledge or about particular issues of right and wrong. The (morally) virtuous individual is traditionally conceived as capable of knowing right from wrong throughout the whole range of circumstances that might arise in his life, but I very much doubt whether anyone has ever possessed such general knowledge. Thus in the ancient world—in Greece, in Rome, among the Church Fathers—no one ever wrote or is known to have spoken against slavery as such. (There were occasional discussions of the abuses and justifications of slavery, but as far as we know, it was never held that slavery was simply wrong.)[1] And the very fact that slavery was never questioned in the

[1] See M. I. Finley, *Ancient Slavery and Modern Ideology*, London: Chatto and Windus, 1980, pp. 120 f. Finley mentions only Euripides as a possible exception to the above generalization. For further discussion of the 'epistemology of virtue' see my 'Is Virtue Possible?', *Analysis* 42, 1982, pp. 70–76.

ancient world, but was simply accepted as an inevitable, indeed as a 'natural', fact should make us question whether anyone today can answer all the important moral questions that confront us.

Some day our descendants may regard twentieth-century morality and moral philosophy as having been as hopelessly wrong-headed on certain issues (vegetarianism? children's rights? patients' rights?) as we regard ancient views on slavery. And with as much reason. So perhaps moral virtue as traditionally understood is to be attained, if at all, only at the historical limit of human cultural endeavour, in a long run that no individual may ever encompass. But that does not undercut the possibility of objective morality nor deprive us of the right to make particular moral judgements (and I have, of course, made plenty of them here).

The notion of a personal (or life) good is not encumbered by the epistemic requirements we find in the traditional idea of a (fully) morally virtuous individual, and so various human goods (though not, if it is a good, the good of being morally virtuous) are presumably available here and now and almost always have been. But in speaking of various goods singly and in indicating something about how they (and their timing) contribute to the overall goodness of lives, I have said nothing about *the* good life and this has been deliberate. Many recent philosophers have questioned whether any single kind of life can be considered best,[2] and the previous two chapters should do something to support this point of view through their attempt to show that ethical considerations need not constrain what counts as in an individual's interest or to his advantage. For if 'good life' refers to personal well-being rather than (just) to admirability or virtue, then our discussion raises the possibility that even the lives of the unjust (or addictive or sadistic) may represent forms of personal well-being (for those who have those lives), and the question is at least left open whether in these terms such lives may not be just as good as those we find less odious. And

[2] I have particularly in mind the views of John Rawls and Isaiah Berlin.

once suspicions about the ideal of *autarkeia* have gained a foothold, we may also be in a better position to resist the idea that the detached life of reason, of contemplation, must set the standard for our own.

Index